HLA and Disease: A Comprehensive Review

Author

William E. Braun, M.D.

Director, Histocompatibility Laboratory
Chief, Medical Renal Transplantation Service
The Cleveland Clinic
Cleveland, Ohio

CRC PRESS, INC.
Boca Raton, Florida 33431

Library of Congress Cataloging in Publication Data

Braun, William E
HLA and disease.

Bibliography: p.
Includes index.
1. HLA histocompatibility antigens. 2. Diseases — Causes and theories of causation. 3. Man—Constitution. I. Title. [DNLM: 1. HL-A antigens. QW573 B825h]
QR184.3.B72 616.07'1 79-4182
ISBN 0-8493-5795-0

This book represents information obtained from authentic and highly regarded sources. Reprinted material is quoted with permission, and sources are indicated. A wide variety of references are listed. Every reasonable effort has been made to give reliable data and information, but the author and the publisher cannot assume responsibility for the validity of all materials or for the consequences of their use.

International Standard Book Number 0-8493-5795-0

Library of Congress Card Number 79-4182
Printed in the United States

THE AUTHOR

William E. Braun, M.D., has been Director of the Histocompatibility Laboratory at the Cleveland Clinic since 1968, Chief of the Medical Renal Transplantation Service since 1972, and author of more than 70 articles in these areas. As the applications of HLA testing expanded from an exclusive use for renal transplantation to disease associations, so too have his investigative efforts.

He was the first president of the American Association for Clinical Histocompatibility Testing and also has served on the Transplantation and Immunology Branch Advisory Board of NIAID, the Policy Board of the Kidney Transplant Histocompatibility Study of the NIH, and the American Medical Association Committee on Transfusion and Transplantation. He is a member of the American Federation for Clinical Research, American Society of Nephrology, the Transplantation Society, and the American Association for Clinical Histocompatibility Testing. Dr. Braun is certified by the American Board of Internal Medicine and the American Board of Nephrology.

ACKNOWLEDGMENTS

We would like to express our appreciation to Dr. Frances Ward of Duke University for reviewing the manuscript and to Mrs. Nora Kancelbaum for her expert secretarial and editing assistance in its preparation.

TABLE OF CONTENTS

Chapter 1

INTRODUCTION TO THE HLA ANTIGEN SYSTEM

I. HISTORY

In 1958, Dausset's discovery of an antibody detecting a human leukocyte antigen (Mac) (HLA-A2)[1] introduced the science of human histocompatibility testing. Leukocyte antigens, initially tested only for organ transplantations[2-4] and later blood transfusions,[5-7] are now known to have a strong relationship to immune responsiveness and disease susceptibility.[8-11] The current and future clinical usefulness of HLA testing is based on several major scientific observations and technologic advances such as the discovery of parous females as an excellent source of antibody,[12] the miniaturization and standardization of the cytotoxicity technique,[13] the establishment of the Serum Bank at the National Institutes of Health as a repository for quality typing reagents,[14] the ability to freeze lymphocytes for reference cell typing, the conversion of the mixed lymphocyte culture to a unidirectional test[15] and its reduction to a micromethod,[16] the exploration of cellularly defined antigens by homozygous typing cells (HTC)[17] and primed lymphocyte testing (PLT),[18] and the rapidly developing identification of new leukocyte antigen systems by serologic reagents for B-cell antigens.[19,20]

Each of the international workshops for the study of histocompatibility antigens has established a milestone of progress. The first workshop was organized by Amos of Duke University in order to compare a variety of evolving techniques used for identifying leukocyte antigens. In 1965, participants in the second workshop organized by van Rood in Leiden, Holland tested with their own techniques and sera individuals of the "Leiden panel" and found similarly reacting sera and closely related antigens. For the Turin workshop in 1967, Dr. Ceppellini selected Italian families for collaborative study and showed that leukocyte antigens were controlled by a single chromosome region and were highly polymorphic. The fourth workshop, conducted by Dr. Paul Terasaki in Palm Springs, California in 1970, again utilized information on family typings provided by each investigator working in his own laboratory with a standardized technique and panel of typing sera.

As the number of accepted antigens increased, it was established in 1970 that there were at least two segregant series of antigens, each behaving as if the corresponding antigens were controlled by a multiple allelic series at a single complex genetic locus or at closely linked loci. But already, cracks were appearing in the foundation of some of the previously accepted antigens such as A3, B5, and B7, and soon the complexity of cross-reacting antigens and "splits" began to unfold.[3]

Population studies done throughout the world were the subject of the fifth workshop in Evian, France conducted by Dr. Jean Dausset in 1972. It showed the strong influence of race on the frequency of histocompatibility antigens.

Under the direction of Dr. Fleming Kissmeyer-Nielsen, the sixth international workshop in Aarhus, Denmark investigated not only the serologically defined (SD) antigens of the A, B, and C loci but also the lymphocyte-defined (LD) antigens of the D locus defined by homozygous typing cells (HTC) in unidirectional mixed lymphocyte culture. The 1977 workshop in Oxford, England, organized by Drs. Walter and Julia Bodmer, focused primarily on B lymphocyte antigens (DR antigens) that may be analogous to Ia antigens studied in animals. Details of these systems in mice have been recently reviewed[21] and are discussed later.

II. NOMENCLATURE

Following the 1977 workshop, the World Health Organization HLA Nomenclature Committee provided new terminology for all of the 59 SD antigens of the A, B, and C series, the 11 LD antigens of the newly described D series, and the 7 new DR antigens. The region or system designation for the histocompatibility antigens remains HLA, formerly written HL-A. The original interpretation of HLA was human leukocyte locus A[22] but is now considered more appropriately as the human lymphocyte antigen system, since antigens of all four of the known loci of this system have been detected on lymphocytes. The five loci thus far defined within the HLA system have been given the designations A, B, C, D, and DR. The A, B, C, and DR series of antigens are serologically defined and thus sometimes designated SD, whereas the D locus antigens, which initially and primarily were detectable by mixed lymphocyte culture reactions, have been designated as cellularly defined (CD) or lymphocyte defined (LD).[23] Antigens of each of these loci that have official World Health Organization acceptance are now designated as HLA-A, HLA-B, HLA-C, HLA-D, and HLA-DR with the appropriate arabic number behind it for the antigen specificity. If an antigen is only provisionally accepted, a "w" is inserted between the locus letter and the antigen number, e.g., HLA-Bw22.

For reference to the older literature, it should be recalled that the A locus was previously known as the first locus, and before that, the LA locus; the B locus was previously known as the second or the Four locus; the C locus was previously known as the third locus or as the AJ locus; and the D locus was known as the mixed lymphocyte culture (MLC), the mixed lymphocyte reaction (MLR), the LD, and briefly as the "a" locus. The DR antigens, which initially were thought to be simply the serologic equivalents of HTC-derived D antigens are now felt to represent the gene products of a distinct, nearby locus, DR.

An example of a genotype for an individual written in terms of the two haplotypes for that individual would be as follows: HLA-A2, B12, Cw5, DRw4/HLA-A3, Bw22, Cw1, DRw2 (Table 1).

If there is an undetected antigen in a family study, then the unknown specificity may be designated with the capital letter designating the locus to which it belongs in conjunction with a hyphen afterwards indicating the blank or undesignated antigen.

A complete listing of all the recognized HLA specificities is given in Table 1, and further details of the precise forms for writing genotypes and phenotypes in either extended or abbreviated fashion have been published.[24] More extensive listings of the older nomenclature and equivalents of current antigen specificities are also available.[25]

It should be appreciated that the first human leukocyte antigen was described only about 20 years ago and that the number of accepted antigens has increased rapidly within the last few years. In 1967, there were only six accepted specificities, HLA-A1, 2, 3, and HLA-B5, 7, and 8; whereas in 1970, the number almost doubled to 11 specificities, HLA-A1, 2, 3, 9, 10, 11, and HLA-B5, 7, 8, 12, and 13. The current listing of 77 specificities represents a dramatic increase in the number of antigens to be considered in the use of this system. Part of the increase in the number of these antigens is due to "splits" of previously intact specificities such as A9 splitting into Aw23 and Aw24, while others represent new antigens, e.g., B37. Still there are serologic similarities extending beyond single antigens so that cross-reacting groups (CREGS) are a well-accepted phenomenon.[3]

Another important set of antigens, not yet as well defined in man, are the so-called immune response antigens detected in the mouse and other species of inbred animals. This topic has been reviewed recently[21] and may be summarized as follows. Two antigen systems, Ir and Ia, are located within the I region located in the major histocom-

TABLE 1

New Nomenclature of the HLA System[a]

Locus A	Locus B	Locus C
A1	B5	Cw1
A2	B7	Cw2
A3	B8	Cw3
A9	B12	Cw4
A10	B13	Cw5
A11	B14	Cw6
A25 (10)[b]	B15	
A26 (10)	B17	
A28	B18	Locus D
A29	B27	
Aw19	B37	Dw1
Aw23 (9)	B40	Dw2
Aw24 (9)	Bw4	Dw3
Aw30	Bw6	Dw4
Aw31	Bw16	Dw5
Aw32	Bw21	Dw6
Aw33	Bw22	Dw7
Aw34	Bw35	Dw8
Aw36	Bw38 (16)	Dw9
Aw43	Bw39 (16)	Dw10
	Bw41	Dw11
	Bw42	
	Bw44 (12)	
	Bw45 (12)	Locus DR
	Bw46	
	Bw47	DRw1
	Bw48	DRw2
	Bw49 (21)	DRw3
	Bw50 (21)	DRw4
	Bw51 (5)	DRw5
	Bw52 (5)	DRw6
	Bw53	DRw7
	Bw54 (22)	

Note: A fully defined haplotype is written in long form, for example, as HLA-A2, HLA-B12, HLA-Cw5, HLA-Dw4, HLA-DRw4; and in short form as A2, B12, Cw5, Dw4, DRw4.

[a] Antigens that are provisionally accepted in the HLA system have the prefix "w." Note that about one half of the A series, two thirds of the B series, and all of the C, D, and DR series antigens are provisionally accepted.

[b] Numbers in parentheses refer to the primary antigen from which the antigen splits were derived.

patibility complex of the mouse (H-2). The Ir designation is for immune response antigens which are predominantly on T lymphocytes and which can be detected by: (1) different degrees of antibody response in an animal following immunization with specific antigens as well as (2) different degrees of susceptibility to viral oncogenesis. Ia stands for I-region-associated antigens that are found predominantly on B lymphocytes and can be serologically determined. The Ir and Ia antigens are located close to the murine histocompatibility antigens and may represent analogues of similar but as yet ill-defined antigens in man which also may be determinants of immune responsiveness and disease susceptibility. The I region antigens are discussed further under Section II.

TABLE 2

The Current HLA Histocompatibility System in Man

Histocompatibility Antigens
(Transplantation Antigens)

	HLA				
Loci	DR	D	B	C	A
Old Terminology	B cell	MLC[a]	2nd	3rd	1st
		MLR	Four	T	LA
		a	—	AJ	—
		LD	—	—	—
Techniques for detection	SD	LD	SD	SD	SD
T lymphocytes	—	—	+	+	+
B lymphocytes	+	+	+	+	+

[a] Because as many as three loci may influence the MLC, one must be cautious in equating stimulation in HLC with nonidentity for D-locus antigens.

TABLE 3

Immune-response, Possible Disease-Susceptibility, and H-2 Histocompatibility Systems in the Mouse[a]

Region	K	I					S	G	D
Subregion	—	I-A	I-B	I-J	I-E	I-C	—	—	—
Loci	H-2K	Ir-1A[b]	Ir-1B	—	Ir-1C	Ir-1C	S,	—	H-2D
		Ia-1	Ia-2	Ia-4	Ia-5	Ia-3	—	—	—
Techniques for detection	SD	Immunization with specific antigens and assay of antibody response; susceptibility to viral oncogenesis; LD[b]			SD		Serum protein and some complement components	SD[c]	SD

[a] References 21 and 71.
[b] Strong MLC stimulation in the mouse is most likely based on differences at Ir-1A, although other loci can also influence the MLC response.
[c] Erythrocyte antigens.

The HLA system of antigens and the immune response antigens of the mouse have been reduced to simplified schemes in Tables 2 and 3, respectively. The previous terminology for these antigen systems, the primary techniques used to detect them, as well as the lymphocyte subtypes on which they are predominantly found are also presented.

III. FUNDAMENTAL FACTS CONCERNING HLA ANTIGENS

The HLA antigens are glycoproteins which can be found in the serum, saliva, and urine of man. The antigens exist on the surface of all tested cells in the human body with the possible exception of the mature erythrocyte and trophoblast. They exist on the reticulocyte but appear to be virtually extinguished on the mature erythrocyte, although some evidence indicates that certain antigens may still be present in small quantities here also.[26] Furthermore, they may possibly be present on the surface of human sperm as single haplotypes. In the renal glomerulus, HLA antigens are located in the cellular fraction but have not been detected in the glomerular basement membrane component.[27] Studies of HLA antigens solubilized with papain from the cell-surface membranes of cultured lymphoblastoid cells have led to the description of the HLA antigen as composed of two noncovalently-bound polypeptide chains, a heavier one with a molecular weight of about 33,000 daltons and a lighter one with a weight of approximately 11,000 daltons.[28] The 11,000-dalton polypeptide is identical to the beta-2-microglobulin protein which in turn exhibits structural homology with the CH3 region of the IgG heavy chain. Molecules of beta-2-microglobulin appear to be more numerous (6×10^7 molecules) than the 33,000-dalton HLA molecules (4.5×10^4) on the cell surface of the human lymphocyte.[29] Although the heavy and the light chain are closely related anatomically on the cell surface of lymphocytes, [29] the genetic control for these two polypeptide chains appears to be different. Genetic control for the heavy chain of the HLA molecule is on the 6th chromosome,[30] whereas that for beta-2-microglobulin is on the 15th chromosome.[31]

The other major lymphocyte antigen system currently under intense investigation is that on B lymphocytes. Biochemical studies on these so-called Ia antigens derived by papain treatment have shown that they also have two noncovalently-bound polypeptide chains, one about 30,000 and the other about 23,000 daltons, but no beta-2-microglobulin.

In order to avoid confusion between the HLA antigen system and other antigen systems unique to a certain cell type, it should be stated here that although lymphocytes, neutrophils, and platelets all have HLA antigens, in addition, lymphocytes have non-HLA antigen systems,[32] neutrophils have three known systems (NA, NB, and 9),[33] and platelets have specific systems (PLA, PLE, and KO). Other cells may have differentiation antigens as well which are characteristic of their own histologic type.

Antibodies to HLA antigens are generally of the IgG class but IgM antibodies have also been reported.[3] The IgM antibodies in particular seem to have a greater reactivity at lower temperatures and comprise some of the autocytotoxins seen in diseases such as lupus erythematosus.[35]

Within the HLA system there is suggestive evidence that certain antigens may be stronger transplantation antigens than others. Both clinical and experimental studies are compatible with the concept that HLA-A2 is a relatively strong histocompatibility antigen[36-39] and conversely, that HLA-A1, 3, and 11 are weaker.[37,40,41] The proximity of the B locus to the D locus has made the B-series antigens appear stronger, a point borne out in several clinical studies though not in all.[2-4,42-45] From the results of immunization studies, the C-series antigens appear to be weak.[46] Skin-graft studies would suggest that D-locus antigens are more important than any of the SD antigens.[47,48] A clinical bone marrow success in which only the D-locus antigens were matched and all the SD antigens were incompatible also support this possibility,[49] although recently, a D-locus incompatible, SD-identical graft has also been successful.[50] Another factor that affects the strength of response to histo*in*compatible antigens is the presence or absence of cross reactivity between donor and recipient antigens, but the direction of the cross reactivity is important.[51]

IV. TECHNIQUES FOR THE DETECTION OF HLA ANTIGENS

The technique used in virtually all major histocompatibility laboratories is the Ter-

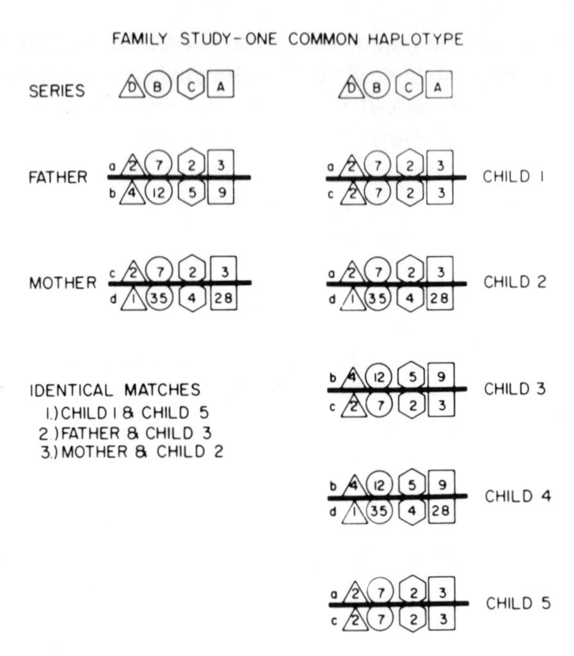

FIGURE 1. The small letters a, b, c, and d denote an entire haplotype. In the family shown, a shared parental haplotype, A3, B7, Cw2, Dw2, permits identical matches between the father and child 3, and between the mother and child 2. Among the children, there is one HLA-identical pair, children 1 and 5, who are homozygous for every antigen including D which makes them useful as LD (HTC) typing cells in MLC. The new DR locus is not shown.

asaki microcytotoxicity technique, [13] but the Amos two-step modification and the fluorochromasia techniques are also popular.[52] The cytotoxicity technique has been used to detect antigens of the A,B, and C series and is being used with a longer incubation period to find serologic reagents capable of identifying on B-cell enriched peripheral lymphocytes and on cells from patients with chronic lymphocytic leukemia antigens of the DR series.

The D-series antigens are defined primarily by mixed lymphocyte culture (MLC) reactions. The MLC technique, which measures the recognition phase of the immune response, has been rendered a unidirectional test, [15] and has been reduced to a micromethod also.[16] It can be used to detect D-series antigens in two ways. First, overall incompatibility or compatibility may be established in unidirectional tests by reactivity or lack of reactivity, respectively, of donor and recipient cells without an attempt to specify which D antigens are involved. Second, the MLC reaction can be used to identify specific D-series antigens by means of homozygous typing cells (HTC) that are homozygous for a D-series antigen.[17] An example of an HTC is shown in the family in Figure 1. Further discussion of LD typing protocols can be found in recent publications.[8,37,53] The D antigens defined by homozygous typing cells (HTC) have been compared to DR antigens identified by B-cell antisera.[22]

Another approach to identifying D-series antigens is the primed lymphocyte test (PLT).[18] After priming lymphocytes to a mismatched D antigen by earlier culture with selected lymphocytes that differ by only a single haplotype, the newly primed lymphocytes will theoretically respond fully only when restimulated by cells with the same D antigen.

Cell-mediated lympholysis (CML) extends the information derived from the mixed lymphocyte culture test and takes it a step further to the effector phase of the immune response. The effector phase is manifested by destruction of target cells and occurs in the absence of antibody. The effector cells require sensitization by D-series antigens but exert their effect against cells with serologically defined differences (i.e., A-, B-, and C-series antigens).[37] Consequently, the specificity of this effector cell cytotoxicity is directed toward the serologically defined antigens present on the stimulating cell and absent on the responding cell and not against the different D antigens of the stimulator cell.

The tests used to detect presensitization to histocompatibility antigens (crossmatching techniques) are now quite numerous and have confusing titles. First, there is the basic antibody-dependent, complement-dependent cytotoxicity crossmatch which may be modified in a number of ways.[54] Modification of the complement-dependent cytotoxicity test has been used for detecting presensitization to B-cell antigens. Next, there are a group of tests which are antibody-dependent but complement-*in*dependent and that reportedly can improve the sensitivity of preformed antibody detection. These tests include the antiglobulin technique,[55] antibody-mediated, cell-dependent immune lympholysis (ABCIL),[56] antibody-dependent, cell-mediated cytotoxicity (ADCC),[57] lymphocyte-antibody lymphocytolytic interaction (LALI),[58] and lymphocyte-dependent antibody (LDA).[59] In addition, antibodies that may or may not be in the HLA system are detected by MLC blocking,[60,61] capillary agglutination,[62] B-cell testing,[63] as well as by those of the techniques noted above which are complement-independent. Finally, products of sensitized lymphocytes such as migration inhibiting factor (MIF)[64] may also be used as an indicator of presensitization. As yet, there are no comprehensive comparative studies of these various techniques. Currently, the antibody-dependent, complement-dependent cytotoxicity test with its modifications is the standard test used for detection of presensitization.

V. GENETICS OF THE HLA SYSTEM

The A, B, C, D, and DR series of HLA antigens are controlled by linked but distinct loci on the sixth chromosome. The gene at each locus has multiple alleles, the products of which are the antigen specificities noted in Table 1. The HLA alleles are codominant. The unit of inheritance of these antigens is designated a haplotype which represents the contribution of one parent to the offspring. A haplotype will have no more than one antigen from each series, currently giving a total of five. The phenotype of an individual who has inherited one haplotype from each parent will have a total of ten antigens, two from each series. An example of a family with all of the antigens detected in each series is shown in Figure 2. In this family, the parents each have two different haplotypes. This is the typical situation in a random population with four different haplotypes occurring in the two parents. Among their children, there is a 25% chance of two of the children being a 2-haplotype identical match, a 25% chance of a 0-haplotype match, and a 50% chance of a 1-haplotype match. The details of this are described in Figure 2.

If the family is one in which the parents share a single haplotype, then in addition to the HLA-identical sibling pair, there is now the possibility of identical parent-child matches as shown in Figure 1. Furthermore, the sharing of a parental haplotype, that may be especially found in the marriages of first cousins, can lead to offspring who are homozygous for all of the antigens of the A, B, C, and D series. The homozygosity of the D-series antigen renders such an individual's cells useful as "LD typing cells" or HTC as described earlier.

When two antigens of different series are associated with a greater degree of frequency than would be expected on the basis of their independent frequencies in the

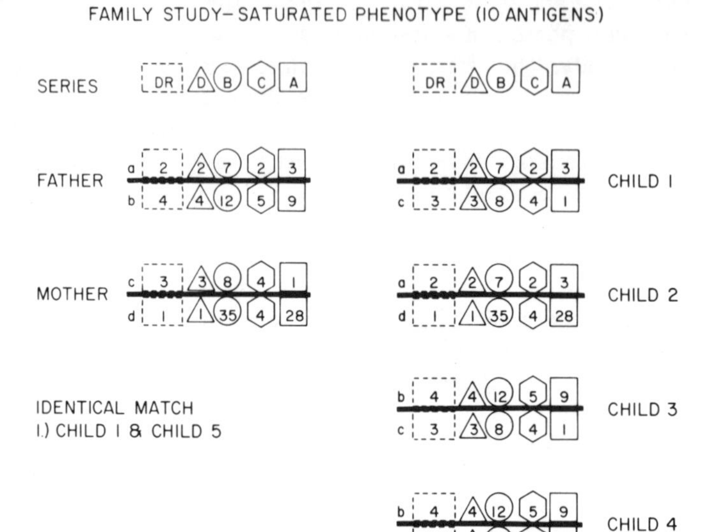

FIGURE 2. The small letters a, b, c, and d denote an entire haplotype. In the family shown, the parents have four different haplotypes. For child 5, there is a 25% chance of a 2-haplotype-identical sibling (child 1), a 25% chance of a 0-haplotype-identical sibling (child 4), and a 50% chance of a 1-haplotype-identical sibling (children 2 and 3). (From Braun, W. E., *Laboratory Medicine,* Vol. 2, Harper & Row, New York, chap. 35. Reproduced by permission of Harper & Row.)

population, the affinity of these antigens is called "linkage disequilibrium." A number of such associations have been shown in Caucasians. Examples of these associations between antigens of the A and B series are A1-B8, A3-B7, A2-B12, and A29-B12; between antigens of the B and C series, B5-Cw1, Bw22-Cw1, B27-Cw2, B40-Cw3, and Bw35-Cw4; between antigens of the B and D series, Bw35-Dw1, B7-Dw2, B8-Dw3, and B15-Dw4; and between antigens of the A, B, and C series, Aw23, B12, and Cw4. The clinical usefulness of linkage disequilibrium lies in the fact that if one is unable to specifically identify a D or DR series antigen, one increases the probability of identifying it by finding the B- or the D-series antigen with which the missing antigen is in linkage disequilibrium.

Individual HLA antigen frequencies and the sum of the frequencies of antigens defined in each of these series can vary significantly according to race. For example, in Caucasians, approximately 98% of the antigens in the A series, 90% of those in the B series, and 50% of those in the C, D, and DR series have been identified. However, in Blacks, Orientals, and other races, fewer of the antigens in these series are identified than in the Caucasian population. Therefore, other antigens, in addition to those found in Caucasians, may exist in these other races and still must be identified. An example of individual antigen frequencies changing with race can be seen with A1 and B8; these are among the most frequent antigens found in Caucasians but are not found in Japanese.[3,65] The Fifth International Histocompatibility Workshop was devoted to the study of racial influences on HLA.[66]

Genetic mapping of the histocompatibility loci on the sixth chromosome places the A, B, C, and D loci in a reverse, skip order: D, B, C, A. The DR locus is probably closest to the D locus. Also mapped in this chromosomal region, called the Major

Histocompatibility Complex (MHC), are the genes, which control the levels of C2 and C4; the structural gene for C2; properdin factor B, also known as glycine-rich beta-glucopeptide (Bf); and the Chido and Rodgers blood groups now known to be components of C4. Data concerning the locus for C6 being in the MHC are conflicting.[68,69] The Ia antigens may well be controlled by several genes at loci distributed throughout the MHC and intermingled with the HLA loci. Although not yet formally assigned to this region, immune response genes and genes conferring susceptibility to certain diseases have been placed in the MHC in man. Also closely associated with D-locus gene products is the lymphocyte Fc receptor.[70]

REFERENCES

1. **Dausset, J.**, Iso-leuco-anticorps, *Acta Haematol.*, 20,156,1958.
2. **Advisory Committee to the Renal Transplant Registry**, The twelfth report of the Human Renal Transplant Registry, *JAMA*, 233, 787, 1975.
3. **Amos,D.B. and Ward, F.E.**, Immunogenetics of the HLA system, *Physiol. Rev.*, 55, 206, 1975.
4. **Dausset, J., Hors, J., Busson, M., Festenstein, H., Oliver, R.T.O., Paris, A.M.I., and Sachs, J.A.**, Serologically defined HL-A antigens and long-term survival of cadaver kidney transplants, *N. Engl. J. Med.*, 290, 979, 1974.
5. **Graw, R.G., Jr., Herzig, G., Perry, S., and Henderson, E.S.**, Normal granulocyte transfusion therapy, *N. Engl. J. Med.*, 287, 367, 1972.
6. **Van Rood, J.J.**, The HL-A system. II. Clinical relevance, *Semin. Hematol.*, 11, 253, 1974.
7. **Yankee, R.A., Graff, K.S., Dowling, R., and Henderson, E.S.**, Selection of unrelated compatible platelet donors by lymphocyte HL-A matching, *N. Engl. J. Med.*, 288, 760, 1973.
8. **Ceppellini, R. and van Rood, J.J.**, The HL-A system. I. Genetics and molecular biology, *Semin. Hematol.*, ll, 233, 1974.
9. **Dausset, J. and Hors, J.**, Some contributions of the HL-A complex to the genetics of human diseases, *Transplant. Rev.*, 22, 44, 1975.
10. **Svejgaard, A., Platz, P., Ryder, L.P., Staub Nielsen, L., and Thomsen, M.**, HLA and disease association — a survey, *Transplant. Rev.* 22, 3, 1975.
11. *HLA and Disease,* INSERM, Paris, 1976.
12. **Payne, R. and Rolfs, M.R.**, Fetomaternal leukocyte incompatibility, *J. Clin. Invest.*, 37, 1756, 1958.
13. **Mittal, K.K., Mickey, M.R., Singal, D.P., and Terasaki, P.I.**, Serotyping for homotransplantation. XVIII. Refinement of microdroplet lymphocyte cytotoxicity test, *Transplantation,* 6, 913, 1968.
14. **Brand, D.L., Ray, J.G., Jr., Hare, D.B., Kayhoe, D.E., and McClelland, J.D.**, Preliminary trials toward standardization of leukocyte typing, in *Histocompatibility Testing 1970,* Terasaki, P.I., Ed., Munksgaard, Copenhagen, 1970, 357.
15. **Bach, F.H. and Voynow, N.K.**, One way stimulation in mixed lymphocyte cultures, *Science,* 153, 545, 1966.
16. **Hartzman, R.J., Segall, M., Bach, M.L., and Bach, F.H.**, Histocompatibility matching. VI. Miniaturization of the mixed leukocyte culture test: a preliminary report, *Transplantation,* 11, 268, 1971.
17. **Dupont, B., Jersild, C., Hansen, G.S., Staub Nielsen, L., Thomsen, M., and Svejgaard, A.**, Typing for MLC determinants by means of LD-homozygous and LD-heterozygous test cells, *Transplant. Proc.*, 5, 1543, 1973.
18. **Sheehy, M.J., Sondel, P.M., Bach, M.L., Wank, R., and Bach, F.H.**, HLA LD (Lymphocyte Defined) typing: a rapid assay with primed lymphocytes, *Science,* 188, 1308, 1975.
19. **Ting, A., Mickey, M.R., and Terasaki, P.I.**, B-lymphocyte alloantigens in Caucasians, *J. Exp. Med.*, 143, 981, 1976.
20. **Walford, R.L., Gossett, T., Troup, G.M., Gatti, R.A., Mittal, K.K., Robins, A., Ferrara, G.B., and Zeller, E.**, The Merrit alloantigenic system of human B lymphocytes: evidence for thirteen possible factors including one six-member segregant series, *J. Immunol.*, 116, 1704, 1976.
21. **Shreffler, D.C. and David, C.S.**, The H-2 major histocompatibility complex and the I immune response region: genetic variation, function, and organization, *Adv. Immunol.*, 20, 125, 1975.
22. **Kissmeyer-Nielsen, F. and Thorsby, E.**, Human transplantation antigens, *Transplant. Rev.*, 4, 5, 1970.

23. **Whitsett, C., Lee, T.D., Duquesnoy, R., Noreen, H., Hansen, J.A., Annen, K., Reinsmoen, N., Long, M.A., Yunis, E.J., and Dupont, B.**, B-lymphocyte alloantigens and HLA-D determinants in a North American white population, *Transplant. Proc.*, 9, 425, 1977.
24. New nomenclature for the HLA system, *J. Immunol.*, 116, 573, 1976.
25. Table of equivalent nomenclature, in *Histocompatibility Testing 1972*, Dausset, J. and Colombani, J., Eds., Williams & Wilkins, Baltimore, 1972, 7.
26. **Doughty, R.W., Goodier, S.R., and Gelsthorpe, K.**, Further evidence for HL-A antigens present on adult peripheral red blood cells, *Tissue Antigens*, 3, 189, 1973.
27. **Braun, W.E. and Murphy, J.J.**, HL-A antigens of the human renal glomerulus, *Transplant. Proc.*, 3, 137, 1971.
28. **Tanigaki, N. and Pressman, D.**, The basic structure and the antigenic characteristics of HL-A antigens, *Transplant. Rev.*, 21, 15, 1974.
29. **Solheim, B.G.**, Association between the β-2-m/HL-A molecule and membrane structures responsible for lymphocyte activation, *Transplant. Rev.*, 21, 35, 1974.
30. **Van Someren, H., Westerveld, A., Hagemeijer, A., Mees, J.R., Khan, P.M., and Zaalberg, O.B.**, Human antigen and enzyme markers in man-Chinese hamster somatic cell hybrids: evidence for synteny between the HL-A, PGM_3, ME, and IPO-B loci, *Proc. Natl. Acad. Sci. U.S.A.*, 71, 962, 1974.
31. **Goodfellow, P.N., Jones, E.A., Van Heyningen, V., Solomon, E., Bobrow, M., Miggiano, V., and Bodmer, W. F.**, The β_2-microglobulin gene is on chromosome 15 and not in the HLA region, *Nature* (London), 254, 267, 1975.
32. **Walford, R.L.**, The isoantigenic systems of human leukocytes: medical and biological significance, *Ser. Haematol.*, 2(2), 5, 1969.
33. **Lalezari, P. and Radel, E.**, Neutrophil-specific antigens: immunology and clinical significance, *Semin. Hematol.*, 11, 281, 1974.
34. **Svejgaard, A.**, Iso-antigenic systems of human blood platelets. A survey, *Ser. Haematol.*, 2(3), 5, 1969.
35. **Terasaki, P.I., Mottironi, V.D., and Barnett, E.V.**, Cytotoxins in disease; autocytotoxins in lupus, *N. Engl. J. Med.*, 283, 724, 1970.
36. **Dausset, J., Rapaport, F.T., Legrand, L., Colombani, J., and Marcelli-Barge, A.**, Skin allograft survival in 238 human subjects, in *Histocompatibility Testing 1970*, Terasaki, P.I., Ed., Munksgaard, Copenhagen, 1970, 381.
37. **Eijsvoogel, V.P.**, The cellular recognition in vitro of antigens related to human histocompatibility, *Semin. Hematol.*, 11, 305, 1974.
38. **Johnson, A.H., Amos, D.B., and Noreen, H.**, Strong mixed lymphocyte reaction associated with the LA or first locus HLA, *Transplantation*, 20, 291, 1975.
39. **Opelz, G., Mickey, M.R., and Terasaki, P.I.**, HL-A and kidney transplants: reexamination, *Transplantation*, 17, 371, 1974.
40. **Braun, W.E., Straffon, R.A., Nakamoto, S., Popowniak, K.L., Gifford, R.W., Kuruvila, C., and Zachary, A.A.**, Mismatched HL-A haplotypes with antigens HL-A1, 3, and 11 associated with excellent renal allograft function, *Transplantation*, 15, 86, 1973.
41. **Oh, J.H. and Maclean, L.D.**, Comparative immunogenicity of HLA antigens: a study in primiparas, *Tissue Antigens*, 5, 33, 1975.
42. **Oliver, R.T.D., Sachs, J.A., and Festenstein, H.**, A collaborative scheme for tissue typing and matching in renal transplantation. VI. Clinical relevance of HLA matching in 349 cadaver renal transplants, *Transplant. Proc.*, 5, 245, 1973.
43. Scandiatransplant Report: HLA matching and kidney graft survival, *Lancet*, 1, 240, 1975.
44. **Van Hooff, J.P., Schippers, H.M.A., van der Steen, G.J., and van Rood, J.J.**, Efficacy of HL-A matching in Eurotransplant, *Lancet*, 2, 1385, 1972.
45. **Van Hooff, J.P., Schippers, H.M.A., Hendriks, G.F.J., and van Rood, J.J.**, Influence of possible HL-A haploidentity on renal-graft survival in Eurotransplant, *Lancet*, 1, 1130, 1974.
46. **Ferrara, G.B., Tosi, R.M., Longo, A., Azzolina, G., and Carminati, G.**, Low immunogenicity of the third HLA series, *Transplantation*, 20, 340, 1975.
47. **Bach, F.H., and Amos, D.B.**, Hu-1: major histocompatibility locus in man, *Science*, 156, 1506, 1967.
48. **Van Rood, J.J., Koch, C.T., van Hooff, J.P., van Leeuwen, A., van den Tweel, J.G., Frederiks, E., Schippers, H.M.A., Hendriks, G., and van der Steen, G.J.**, Graft survival in unrelated donor-recipient pairs matched for MLC and HL-A, *Transplant. Proc.*, 5, 409, 1973.
49. **Copenhagen Study Group of Immunodeficiencies**, Bone-marrow transplantation from an HLA non-identical but mixed lymphocyte culture-identical donor, *Lancet*, 1, 1146, 1973.
50. **Feig, S.A., Opelz, G., Winter, H.S., Falk, P.M., Neerhout, R.C., Sparkes, R., and Gale, R.P.**, Successful bone marrow transplantation against mixed lymphocyte culture barrier, *Blood*, 48, 385, 1976.
51. **Colombani, J., Colombani, M., Degos, L., Terrier, E., Gaudy, Y., and Dastot, H.**, Effect of cross reactions on HL-A antigen immunogenicity, *Tissue Antigens*, 4, 136, 1974.

52. **Amos, D.B.**, Cytotoxicity testing, in *Manual of Tissue Typing Techniques,* Publication No. (NIH) 75-545, Ray, J.G., Hare, D.B., Pedersen, P.D., and Mullally, D.I., Eds., Department of Health, Education and Welfare, Bethesda, 1975, 23.
53. **Bodmer, W.F.**, Histocompatibility testing international, *Nature* (London), 256, 696, 1975.
54. **Braun, W.E.**, The new serology of histocompatibility testing and its significance in human renal transplantation, *Urol. Clin. N. Am.* 3, 503, 1976.
55. **Johnson, A.H., Rossen, R.D., and Butler, W.T.**, Detection of alloantigens using a sensitive antiglobulin microcytotoxicity test: identification of low levels of pre-formed antibodies in accelerated allograft rejection, *Tissue Antigens,* 2, 215, 1972.
56. **Kovithavongs, T., Olson, L.A., Schlaut, J.W., McConnachie, P.R., and Dossetor, J.B.**, Immunity to tissue sensitisation as detected by the ABCIL system. II. Hemodialysis patients, *Transplantation,* 18, 108, 1974.
57. **Lightbody, J.J. and Rosenberg, J.C.**, Antibody-dependent cell-mediated cytotoxicity in prospective kidney transplant recipients, *J. Immunol.,* 112, 890, 1974.
58. **Trinchieri, G., DeMarchi, M., Mayr, W., Savi, M., and Ceppellini, R.**, Lymphocyte antibody lymphocytolytic interaction (LALI) with special emphasis on HL-A, *Transplant. Proc.,* 5, 1631, 1973.
59. **Ting, A. and Terasaki, P.I.**, Influence of lymphocyte-dependent antibodies on human kidney transplants, *Transplantation,* 18, 371, 1974.
60. **Sengar, D.P.S., Opelz, G., and Terasaki, P.I.**, Suppression of mixed lymphocyte response by plasma from hemodialysis patients, *Tissue Antigens,* 3, 22, 1973.
61. **Suciu-Foca, N., Buda, J.A., Thiem, T., Almojera, P., and Reemstma, K.B.**, Inhibition of mixed leukocyte culture reactivity by sera from hemodialysis patients and transplant recipients as an indicator of isoimmunization, *Lab. Invest.,* 31, 1, 1974.
62. **Thompson, J.S., Severson, C.D., Coppleson, L.W., and Stokes, G.**, Leukocyte capillary agglutination: demonstration of additional leukocyte antibodies in cytotoxically "monospecific" antisera, in *Histocompatibility Testing 1970,* Terasaki, P.I., Ed., Munksgaard, Copenhagen, 1970, 587.
63. **Ettenger, R., Fine, R., and Terasaki, P.I.**, The presence of anti-B cell activity in renal allograft recipients, *Proc. American Society of Nephrology,* American Society of Nephrology, 1975, 100.
64. **Turnipseed, W.D. and Cerilli, J.**, An immunological study of renal allograft rejection using the direct macrophage inhibition test, *Transplantation,* 20, 414, 1975.
65. **Mittal, K.K., Hasegawa, T., Ting, A., Mickey, M.R., and Terasaki, P.I.**, Genetic variation in the HL-A system between Ainus, Japanese, and Caucasians, in *Histocompatibility Testing 1972,* Dausset, J. and Colombani, J., Eds., Munksgaard, Copenhagen, 1973, 187.
66. **Dausset, J. and Colombani, J.**, Eds., *Histocompatibility Testing 1972,* Munksgaard, Copenhagen, 1973.
67. **Bodmer, W.F.**, Report of the committee on the genetic constitution of chromosome 6, in *Human Gene Mapping 3,* Bergsma, D., Ed., S. Karger, New York, 1976, 24.
68. **Carpenter, C.B.**, personal communication, 1976.
69. **Mittal, K.K., Wolski, K.P., Lim, D., Gewurz, A., Gewurz, H., and Schmid, F.R.**, Genetic independence between the HL-A system and deficits in the first and sixth components of complement, *Tissue Antigens,* 7, 97, 1976.
70. **Solheim, B.G., Thorsby, E., and Moller, E.**, Inhibition of the Fc receptor of human lymphoid cells by antisera-recognizing determinants of the HLA system, *J. Exp. Med.,* 143, 1568, 1976.
71. **McDevitt, H. O.**, Conclusions and prospects, in *HLA and Disease,* Dausset, J. and Svejgaard, A., Eds., Munksgaard, Copenhagen, 1977, 314.

Chapter 2

ANIMAL STUDIES AS THE BASIS OF THE RELATIONSHIP OF DISEASE SUSCEPTIBILITY AND IMMUNE RESPONSIVENESS WITH HISTOCOMPATIBILITY, IR, AND IA ANTIGENS

The relationship of human histocompatibility antigens (HLA) with immune responsiveness (IR) and disease susceptibility (DS) has its basis in studies of inbred strains of mice and guinea pigs.[1,2] The most extensive studies have been done in mice. Lilly first found that a gene which conferred resistance to Gross leukemia virus (Rgv-1) appeared to be linked to the H-2 histocompatibility complex.[3] Similarly, an H-2 association was found for Friend, B/T, and radiation leukemia virus, mammary tumor virus, lymphocytic choriomeningitis (LCM), and autoimmune thyroiditis.[1,4,5] Recently, Smith has reported that there are H-2 related differences in the spontaneous occurrence of tumors in inbred mice 9 months or more in age.[6] This first area of disease susceptibility in the form of autoimmunity, virus infection, and viral oncogenesis being related to histocompatibility antigens was expanded into a second area by numerous studies which showed that the immune responsiveness to approximately 20 different natural and synthetic antigens was controlled by genes linked to H-2 histocompatibility genes.[1]

In mice, the major histocompatibility complex (MHC) is located on the 17th chromosome and consists of five major regions: K, I, S, G, and D (Section I, Table 2). The H-2K locus appears analogous to the human HLA-B locus, and the H-2D to the HLA-A locus. The gene products of these loci function as transplantation antigens, weak stimulators in MLC, targets in cytotoxicity, and possibly as a means of immune surveillance. The S-region genes control the production of certain serum proteins including complement and therefore seem related to the complement component genes mapped in the human major histocompatibility complex (MHC). It is the I region that has unfolded with the greatest complexity and has the genes that primarily regulate the immune response to thymus-dependent antigens, antigens that stimulate T cells in the MLC and GVH reaction, and B cell production of Ia antibodies.[7] Recently, an I region locus has been found that controls surface determinants on suppressor T cells.[8] The major MLC locus analogous to the D locus in man is located in the I region in the mouse.

Within the I region there are sets of marker loci: Ir and Ia.[1,9] The Ir designation stands for immune response antigens that appear to be expressed predominantly on T lymphocytes and are detected by different levels of antibody response to specific antigens as well as the degree of susceptibility to viral oncogenesis.[1] Ia refers to I-region associated antigens that are predominantly on B lymphocytes and are serologically determined. Within the I region are the subregions 1-A, 1-B, 1-J, 1-E, and 1-C, in which the Ir- and Ia-marker loci are Ir-1A and Ia-1; Ir-1B and Ia-2; Ia-4, Ir-1C and Ia-5; and Ia-3, respectively (Table 2). Ir-1A is located near the K end of the MHC, and Ia-3 toward the D end. The Ir-1A gene has the strongest, but not the sole, control of MLC reactivity and thus resembles the HLA-D locus of man. Rgv-1 may be an Ir gene located near the K end of the I region and therefore in or near the Ia subregion. A potentially important locus recently described in mice is the Ia-4 locus of the 1-J subregion that controls suppressor cell activity and unlike other Ia loci, has its gene products expressed predominantly if not exclusively on T cells.[8] Unfortunately, no comparable locus has yet been defined in man, though such a locus probably does exist in the MHC and could modify immune responsiveness. The immune responses to natural and synthetic antigens have been mapped with several of the Ia loci.

Until very recently, it was assumed that only one histocompatibility-linked Ir gene was required to control the immune response to a single antigen. But it has been found that the immune response in mice to certain synthetic antigens is controlled by two dominant histocompatibility-linked Ir genes.[10] The response is generally greater in the *cis* than in the *trans* position. McDevitt has expanded this scheme to include five such systems that require a gene in the A subregion and the C subregion complementing each other in the *cis* or *trans* position to give an immune response.[9] The parallel of this in man would be the requirement for one haplotype with both Ir genes or two haplotypes each with an Ir gene probably marked by D-locus antigens. Thus, from elaborate experiments in animals, the evidence that two Ir genes are necessary for the response to an antigen is compatible with the findings in human diseases that suggest the necessity of at least two Ir genes for disease susceptibility.

REFERENCES

1. **Shreffler, D.C. and David, C.S.,** The H-2 major histocompatibility complex and the I immune response region: genetic variation, function, and organization, *Adv. Immunol.,* 20, 125, 1975.
2. **Shevach, E.M., Paul, W.E., and Green, I.,** Histocompatibility-linked immune response gene function in guinea pigs, *J. Exp. Med.,* 136, 1207, 1972.
3. **Lilly, F.,** The histocompatibility-2 locus and susceptibility to tumor induction, *Natl. Cancer Inst. Monogr.,* 22, 631, 1966.
4. **Vladutiu, A.O. and Rose, N.R.,** Autoimmune murine thyroiditis. Relation to histocompatibility (H-2) type, *Science,* 174, 1137, 1971.
5. **Oldstone, M.D., Dixon, F.J., Mitchell, G.F., and McDevitt, H.O.,** Histocompatibility-linked genetic control of disease susceptibility. Murine lymphocytic choriomeningitis virus infection, *J. Exp. Med.,* 137, 1201, 1973.
6. **Smith, G.S., Mickey, M.R., and Walford, R.L.,** Influence of the H-2 system upon lifespan and spontaneous cancer incidence in congenic mice, in *Genetic Effects on Aging,* Harrison, D.E., Ed., The National Foundation, White Plains, in press, 1978.
7. **Klein, J., Geib, R., Chiang, C., and Hamptfeld, V.,** Histocompatibility antigens controlled by the I region of the murine H-2 complex. I. Mapping of H-2A and H-2C loci, *J. Exp. Med.,* 143, 1439, 1976.
8. **Murphy, D.B., Herzenberg, L.A., Okumma, K., Herzenberg, L.A., and McDevitt, H.O.,** A new I subregion (I-J) marked by a locus (Ia-4) controlling surface determinants on suppressor T lymphocytes, *J. Exp. Med.,* 144, 699, 1976.
9. **McDevitt, H. O.,** Conclusions and prospects, in *HLA and Disease,* Dausset, J. and Svejgaard, A., Eds., Munksgaard, Copenhagen, 1977, 314.
10. **Dorf, M. E., Maurer, P.H., Merryman, C.F., and Benacerraf, B.,** Inclusion group systems and cis-trans effects in responses controlled by the two complementing Ir-GLa genes, *J. Exp. Med.,* 143, 889, 1976.

Chapter 3

POSSIBLE MECHANISMS OF HLA AND DISEASE ASSOCIATIONS: SEX PREVALENCE

Perhaps the most important point to make in considering the several mechanisms which may be responsible for histocompatibility antigens being associated with diseases is that different diseases may have different mechanisms and that a disease might have one initiating and another sustaining mechanism. Furthermore, some of these mechanisms blend into one another, e.g., virus receptor and altered self-hypothesis.

The major mechanisms proposed for HLA and disease association are as follows. First, there may be molecular mimicry, or cross-reactivity, between a pathogen and a histocompatibility or disease-association antigen so that normal immune surveillance mechanisms fail to eliminate the pathogen because it resembles a "self" antigen. There is preliminary and rather tenuous evidence for this mechanism in AS where cross-reactivity between B27 and Klebsiella antibodies has been demonstrated.[1] Other indications of cross-reactivity include those between HLA antigens and the streptococcal M-antigen that remains controversial, [2] between transplantation antigens and streptococci,[3] and between allogeneic and autologous platelets in posttransfusion purpura.[4]

Second, a form of molecular mimicry which is nonimmunologic and actually more like competitive inhibition has been hypothesized by Svejgaard.[5] His concept proposes that "if an HL-A antigen has some incidental resemblance to the binding site of a cell-surface receptor molecule for a given hormone, there could be competition between the receptor and the HL-A antigen molecule." The wide tissue distribution of HLA antigens could offer significant interference with ligand (hormone)-receptor interactions despite the stronger affinity between the latter two. The H-2 control of serum testosterone levels and sensitivity of target organs to testosterone are two pieces of evidence presented in support of the hypothesis.

Third, the antigen in question may actually be a receptor for a microbial pathogen or other pathogenic substance. There is no direct evidence for this mechanism, and there is considerable negative evidence from work with measles, lymphocytic choriomeningitis (LCM), and vaccinia virus.[6] The conditions in which there has been some suggestion of this mechanism have been (1) AS with B27 as the presumptive microbial receptor because of its greater than 90% occurrence with AS in Caucasians, (2) gluten sensitive enteropathy because of the response of B8-positive patients to alpha gliadin[7] (see Gluten Sensitive Enteropathy [Chapter 7, Section VII.I]), and (3) vesicular stomatitis virus (VSV)-infected cells because of a 50% decrease in HLA concentration.[6,7] However, in the latter case, the effect of VSV on HLA is most likely based on inhibition of protein synthesis.[6]

A fourth possibility focuses on the histocompatibility restriction for lymphocytes cytotoxic to virus-infected target cells. The major work leading to this concept has been done by Doherty and Zinkernagel who found that cytotoxic T cells from mice acutely infected with LCMV interact only with H-2 compatible virus-infected cells.[8] That this was not an isolated phenomenon was shown by similar H-2 restrictions for T-cell helper activity and for lymphocytes cytotoxic to trinitrophenylated lymphocytes, minor histocompatibility antigens, male y antigen, and numerous viruses (Rous sarcoma, ectromelia, vaccinia, parainfluenza, murine sarcoma, and SV40).

Explanations for the H-2 restriction phenomenon include the dual recognition (or "intimacy") and the altered self (or "interaction") hypotheses. The former theory requires matching of H-2 determinants on the effector T cell and target cell in order for recognition of the target cell to occur. The altered self-hypothesis, which is more likely, proposes that target cells homozygous at the H-2 gene complex express at least two distinct virus-associated antigens (four in heterozygotes). These structures are recognized by different clones of immune T cells generated in virus-infected mice of the same H-2 type. Each antigen is operationally an "interaction antigen" partly coded for by the viral genome and partly by genes mapping at either H-2K or H-2D. The alteration of the cell surface might occur by a conformational change, a hidden or neoantigen, a complex formed between antigen and virus, a biochemical alteration in the antigen, or derepression of other histocompatibility antigens.

Consequently, the heterozygous state of an individual and the polymorphic state of the entire histocompatibility system in a population would tend to be protective mechanisms whereby a broad diversity of viruses and possibly other infectious agents can be successfully repelled by H-restricted cytotoxic lymphocytes.

A fifth and highly promising mechanism is linkage and linkage disequilibrium between the HLA loci and Ir loci for disease susceptibility. The vast majority of such disease associations are with HLA-B and D-series antigens. Svejgaard has pointed out that because Ir genes are dominant, the lack of an adequate Ir determinant could produce a recessive susceptibility to certain infections, whereas an "autoaggressive" Ir determinant could cause a dominant susceptibility to autoimmune disease.[9] However, the modification or production of disease by suppressor cells controlled by an Ia locus must now be considered.[10]

Almost as a corollary of the linked Ir gene mechanism is the sixth mechanism of linked complement and properdin genes. These factors which mediate many forms of immune injury could be important ancillary factors in disease susceptibility but do not seem sufficient by themselves to account for the total phenomenon.

Finally, HLA antigens may act as differentiation antigens that can alter cell-cell interaction. An example of this would be the D antigens that are present only on B cells. This mechanism rests heavily on the tissue distribution of these antigens, but there is relatively little data on the density, distribution, and kinetics of HLA antigens in different tissues.

The well-known sex prevalence of certain diseases such as AS and Reiter's disease in males and of myasthenia and chronic active hepatitis in females, all of which have strong HLA associations, suggests that some influence on disease may be afforded by the combination of both sex and HLA antigens. How this influence of sex may be effected is uncertain, but several possibilities exist. First of all, there may be genes capable of determining the level of male or female hormones which may be one of the elements in the development of the disease. Second, it may be that cell-surface receptors for sex hormones may lie in a close proximity to certain HLA antigens associated with that disease and thereby influence the way in which that antigen operates in establishing susceptibility to a disease. These two mechanisms are also described by Svejgaard's nonimmunologic competitive binding hypothesis and its supporting evidence.[5] Third, the presence of the HY antigen may enhance the development of male-associated diseases and its absence promote the development of female-associated diseases, possibly through related Ia genes or steric interference.[11]

REFERENCES

1. **Ebringer, A., Cowling, P., Ngwa Suh, N., James, D.C.O. and Ebringer, R.W.,** Crossreactivity between Klebsiella aerogenes species and B27 lymphocyte antigens as an aetiological factor in ankylosing spondylitis, in *HLA and Disease,* INSERM, Paris, 1976, 27.

2. **Hirata, A.A. and Terasaki, P.I.**, Cross-reactions between streptococcal M proteins and human transplantation antigens, *Science*, 168, 1095, 1970.
3. **Rapaport, F.T., Chase, R.M., and Soloway, A.C.**, Transplantation antigen activity of bacterial cells in different animal species and intracellular localization, *Ann. N.Y. Acad. Sci.*, 129, 102, 1966.
4. **Svejgaard, A.**, Iso-antigenic systems of human blood platelets — a survey, *Ser. Haematol.*, 2(3), 5, 1969.
5. **Svejgaard, A. and Ryder, L.P.**, Interaction of HLA molecules with nonimmunological ligands as an explanation of HLA and disease association, *Lancet*, 2, 547, 1976.
6. **Oldstone, M.B., Haspel, M.V., Pellegrino, M.A., Kingsbury, D.T., and Olding, L.**, Histocompatibility complex and virus infection latency and activation, *Transplant. Rev.*, 31, 225, 1976.
7. **Nelson, D.L., Falchuk, Z.M., Kasaida, D., and Strober, W.**, Gluten-sensitive enteropathy: correlation of organ culture behavior with HL-A status, *Clin. Res.*, 23, 254, 1975.
8. **Doherty, P.C., Blanden, R.V., and Zinkernagel, R.M.**, Specificity of virus-immune effector T cells for H-2K or H-2D compatible interactions: implications for H-antigen diversity, *Transplant. Rev.*, 29, 89, 1976.
9. **Svejgaard, A.**, HLA and disease, in *Manual of Clinical Immunology*, Rose, N.R. and Friedman, H., Eds., American Society for Microbiology, Washington, 1976, 841.
10. **Murphy, D.B., Herzenberg, L.A., Okumura, K., Herzenberg, L.A., and McDevitt, H.O.**, A new I subregion (I-J) marked by a locus (Ia-4) controlling surface determinants on suppression T lymphocytes, *J. Exp. Med.*, 144, 699, 1976.
11. **Wachtel, S.S., Koo, G.C., and Boyse, E.A.**, Evolutionary conservation of HY male antigen, *Nature* (London), 254, 270, 1975.

Chapter 4

STATISTICAL AND GENETIC CONSIDERATIONS

The first statistical approach to be applied is usually to test for any significant deviation of an HLA antigen frequency in a particular disease as calculated from a 2 × 2 contingency table by the chi-square test or by Fisher's exact test.[1] When one uses the chi-square test, Yates' continuity correction may be included except when a number of chi-square values are being combined. Recently, Yates' correction has been challenged, and its value is currently uncertain. Fisher's exact test is more appropriate when the total sample size is less than 20 or the total sample size is between 20 and 40 and the smallest expected value is less than 5.

When one examines more than one antigen in a patient population, there is an increased possibility of deviations due to pure chance. For example, if 100 random items were studied in two populations, about 5 items could show differences significant at the 5% level purely by chance. As suggested by Wiener [2] for ABO and reemphasized by Grumet[3] for HLA, this type of error may be corrected by multiplying the P value by the number of antigens tested. Many studies were done examining 20 antigens nonselectively; these studies would have had to achieve a P value of 0.0005 in order to be significant at the 1% level when corrected. Now, studies are being done with nearly 40 HLA antigens, and consequently, uncorrected P values would have to be as low as 0.00025 in order to be significant at the 1% level.

A second aspect of the statistical evaluation of disease association studies is the factor described by Woolf[4] as the incidence ratio, modified by Haldane,[5] and now termed the relative risk (RR).[6] The relative risk is a very straightforward indication of how many more times a disease is likely to occur in individuals possessing a certain antigen as it is to occur in those not possessing that same antigen. The relative risk then is not an estimate of how much more frequent an antigen is in the patient group as compared to the control group. The relative risk is determined by the following equation:

$$\frac{h}{k} \times \frac{K}{H} = \text{relative risk (RR)}$$

where h is the number of patients with the particular antigen, k is the number of patients without the antigen, H is the number of controls with the antigen, and K is the number of controls without the antigen.

A third statistical approach is based on Bayes' theorem. It gives an estimation of the probability of a disease being present based on a prior probability of that disease plus the altered frequency of an HLA antigen, the false negativity, and the false positivity of that antigen. This is discussed further in Chapter 5.

Finally, the assessment of linkage between an HLA locus and a disease locus requires a more complex statistical method based on the logarithm of the odds (lod score) of a given sequence of genetic events occurring if the loci are linked as compared to its occurring if the loci are unlinked. Linkage is considered to exist if a lod score exceeds 3.0. The method for calculating the lod score can be found in current genetics texts.[7]

There are some inherent difficulties with HLA antigen assessments.[6] Because these antigens are in segregant series where they are mutually exclusive to a degree, a change in the frequency of an antigen in one direction causes a change in another antigen or several antigens in the opposite direction. In addition, the phenomenon of linkage disequilibrium, which represents a special affinity of one antigen for inheritance with another antigen of a different segregant series, leads to passive increases or decreases in the antigen frequency that is in linkage disequilibrium. Perhaps the best way to ascertain the significance of an antigen frequency deviation is to specifically examine this single antigen in another sample from the same population. When such a study is done, the resulting P value does not require correction for the number of antigens tested because only one was under scrutiny.

An important genetic consideration is the distinction between association, such as is being described for most of these diseases, and genetic linkage, found in only a few. Association is the nonrandom occurrence of two genetically separate traits in a population. Linkage, on the other hand, is the occurrence of two loci on one chromosome, sufficiently close together so that something less than completely independent assortment takes place. Most association has its basis in mechanisms other than genetic linkage. Consequently, it is erroneous to think that finding an association will be the first step in ultimately establishing linkage. An example given by McKusick is that of the Lutheran blood-group locus and the secretor locus.[8] These loci are known to be closely linked. Yet, in any group of unrelated persons, one Lutheran blood type does not occur more frequently with secretor than with nonsecretor. Using a cumulative study of reports, he further points out that the number of persons with both traits, one trait only, or neither trait are about equal. Thus, linkage was not found in these unrelated individuals, and family studies would be necessary to establish its existence.

The HLA system may offer an example of linkage and association existing together.[9] For example, the genes for the five segregant series of HLA antigens are linked, and some of the gene products (HLA antigens) of the different series show some highly significant associations.[9] However, one might argue that this is linkage disequilibrium and should be considered a special case rather than a typical example of association. The suspicion is, though no formal proof exists, that Ir and/or Ia genes linked to HLA antigens could also show association with nearby genes for the A, B, C or D series of HLA antigens. As McKusick states:

> Before enough generations have passed for a chromosome to be minced up by the process of crossing over, association on the basis of genetic linkage may be observed, but linkage produces no permanent association in the population and most association has its basis in mechanisms other than genetic linkage.[8]

Another facet to be considered is the genetic pattern of a disease. A disease may be classified in its genetic pattern as monogenic, oligogenic, or polygenic, each of which may be influenced by environmental factors and thus have the further designation of multifactorial, implying both a genetic and environmental basis.[9,10] A monogenic disorder is due to the action of one gene at one locus. This type of disorder may be inherited as a dominant trait with expression in homo- and heterozygotes or as a recessive with full penetrance and expression only in homozygotes. Since homozygosity is not a requirement for the production of any of the diseases associated with HLA antigens, it appears that the major pattern of disease susceptibility is that of a dominant trait.[9] However, recent animal studies showing that two complementing genes were necessary for a particular type of immune response raise the possibility of similar mechanisms for disease susceptibility also.[11]

When a few genes are involved, approximately two to four, at loci which are not closely linked, the disorder is termed oligogenic. When more than a few genes are involved at loci that are not closely linked, the disorder is termed polygenic, and most of these also have an environmental component. Diseases such as juvenile diabetes mellitus and psoriasis are believed to be oligogenic diseases.[9] It would now appear that virtually all of the HLA-associated diseases are at least oligogenic and probably all are influenced by environment, though the remote possibility exists that ankylosing spondylitis and its association with B27 might be an example of a monogenic disorder with an environmental component.

The simplest analogy that one may think of in looking at the mode of inheritance of diseases associated with HLA antigens is that of opening a lock with a multinotched key. The several notches on the key would represent the specific genes necessary to create the susceptibility to a particular disease. One or more of these notches may be genes for HLA antigens whereas others may be genes for Ir or Ia antigens. The turning of the key in the lock, a necessary event in the process of opening the lock, might be interpreted as the environmental factor that precipitates the disease. Whether all of these genes act simultaneously or whether some of these genes are in an active state and others in a latent state awaiting derepression is even more speculative.

When a significant association has been found in unrelated patients, family studies with several members having the disease are worthwhile for several reasons. Family studies may permit one to show that an antigen occurring with increased frequency in a disease is an inherited characteristic, to explain the role of certain haplotypes, to evaluate the contribution of HLA to the risk of disease, to do linkage studies, and to evaluate other modifying factors, particularly non-HLA genetic factors. Valuable comments on the performance of family studies have recently been provided by Svejgaard.[9,12]

In general, virtually all of the studies thus far reported have looked for the susceptibility to a disease in association with an increased frequency of a known HLA antigen. Studies showing associations with decreased frequencies of HLA antigens, unknown HLA antigens (blank alleles), or non-HLA antigens require larger numbers and are more difficult to perform.

Decreased frequencies of one or more HLA antigens would imply some form of resistance to the disease. A modified or relative resistance to a disease is suggested by the higher frequency of survivors with certain antigens. For example, A2 is increased in survivors of acute lymphocytic leukemia[13] and Aw19 and B5 in survivors of bronchogenic carcinoma.[14] A more absolute, but still incomplete, resistance is suggested by a decreased frequency of certain antigens in different disease groups. For example, in multiple sclerosis, A2 and B12 are decreased, [15] and in celiac disease, and diabetes mellitus B7 is decreased.[16] Finally, what approached an absolute resistance to a disease, leukemia, has been described by Billing as the presence of the B-cell antigen Group 2 of Terasaki.[17] However, the validity of the existence of Terasaki's Group 2 is in serious doubt at present.

While the distinction between the following broad categories might appear tenuous, it does seem that certain antigens as just described for leukemia, multiple sclerosis, and celiac disease may be relevant to whether a person ever contracts a particular disease (disease-susceptibility antigens), whereas other antigens as noted above for acute lymphocytic leukemia and bronchogenic carcinoma may determine how a person responds once he has a disease (immune response antigens).

In conducting any study of HLA and disease, several obvious but frequently overlooked points should be observed. The clinical disease under study should be a homogeneous one, or at least one in which pathologic, physiologic, immunologic, or clinical cleavage lines exist for analysis of subgroups. Ideally, the study should be prospective in order to avoid selection of a patient sample biased toward survival. The temptation must be avoided to study only a group of institutionalized patients with a disease, especially when evaluating correlations with clinical severity. Of great importance is the control group which should be large; unrelated; of the same geographic and racial background as the experimental group; matched for age, sex, and red-cell blood group if appropriate; and have the same duration of exposure in studies of toxicities.

REFERENCES

1. **Armitage, F.,** *Statistical Methods in Medical Research,* John Wiley & Sons, New York, 1971, 134.
2. **Wiener, A.S.,** Blood groups and disease, *Am. J. Hum. Genet.,* 22, 476, 1970.
3. **Grumet, F.C., Coukell, A., Bodmer, J.G., Bodmer, W.F., and McDevitt, H.O.,** Histocompatibility (HL-A) antigens associated with systemic lupus erythematosus, *N. Engl. J. Med.,* 285, 193, 1971.
4. **Woolf, B.,** On Estimating the relation between blood group and disease, *Ann. Hum. Genet.,* 19, 251, 1955.
5. **Haldane, J.B.S.,** The estimation and significance of the logarithm of a ratio of frequencies, *Ann. Hum. Genet.,* 20, 309, 1956.
6. **Svejgaard, A., Jersild, C., Staub Nielsen, L., and Bodmer, W.F.,** HL-A antigens and disease. Statistical and genetic considerations, *Tissue Antigens,* 4, 95, 1974.
7. **Levitan, M. and Montagu, A.,** *Textbook of Human Genetics,* Oxford University Press, New York, 1971, 389.
8. **McKusick, V.A.,** *Human Genetics,* 2nd ed., Prentiss Hall, Englewood Cliffs, 1969.
9. **Svejgaard, A., Platz, P., Ryder, L.P., Staub Nielsen, L., and Thomsen, M.,** HL-A and disease associations — a survey, *Transplant. Rev.,* 22, 3, 1975.
10. **Carter, C.O.,** Genetics of common disorders, *Br. Med. Bull,* 25, 52, 1969.
11. **Dorf, M.E., Maurer, P.H., Merryman, C.F., and Benacerraf, B.,** Inclusion group systems and *cis-trans* effects in responses controlled by the two complementing *Ir-GlΦ* genes, *J. Exp. Med.,* 143, 889, 1976.
12. **Svejgaard, A.,** HLA and disease, in *Manual of Clinical Immunology,* Rose, N.R. and Friedman, H., Eds., American Society for Microbiology, Washington, 1976. chap. 111.
13. **Rogentine, G. N., Trapani, R.J., Yankee, R.A., and Henderson, E.S.,** HL-A antigens and acute lymphocytic leukemia: The nature of the HL-A2 association, *Tissue Antigens,* 3, 470, 1973.
14. **Rogentine, G.N., Jr., Dellon, A.L. and Chretien, P.B.,** Prolonged disease-free survival in bronchogenic carcinoma associated with HLA Aw19 and B5. A follow-up, in *HLA and Disease,* INSERM, Paris, 1976, 234.
15. **Jersild, C., Dupont, B., Fogh, T., Platz, P.J., and Svejgaard, A.,** Histocompatibility determinants in multiple sclerosis, *Transplant Rev.,* 22, 148, 1975.
16. **Ryder, L.P. and Svejgaard, A.,** Associations between HLA and disease, Report from the HLA and Disease Registry of Copenhagen, 1976.
17. **Billing, R.J., Terasaki, P.I., Honig, R., and Peterson, P.,** The absence of B-cell antigen B2 from leukaemia cells and lymphoblastoid cell lines, *Lancet,* 1, 1365, 1976.

Chapter 5

CLINICAL USE OF HLA TYPING

HLA typing may be useful in the diagnosis and prognosis of certain diseases. It can in some instances enhance one's clinical judgment in working through difficult differential diagnoses and may eventually reorient disease categories according to a common pathogenetic mechanism marked by HLA, Ir, or Ia antigens.

When using HLA antigens as an aid in the diagnosis of disease, one must consider the percent false positivity and percent false negativity of any HLA antigen used in this context. For example, the very best test that can be provided by HLA typing is that of B27 for ankylosing spondylitis (AS). Because approximately 10% of individuals with AS lack the B27 antigen, the test has a 10% false negative frequency. On the other hand, since approximately 8% of the normal population has B27, there is an 8% frequency of false positive results. Such a low incidence of false negative and false positive results does not exist for any other HLA antigen associated with a disease. The fact that B27 is not as closely associated with AS in Blacks as in Caucasians should also be appreciated. In Blacks, B27 has a lower false positive rate of about 3% but a much higher false negative rate of about 50%.

Another measurement of the improvement in diagnosis by means of B27 in doubtful cases of AS can be obtained by the use of Bayes' theorem for calculating the a posteriori chances of AS according to the likelihood of the diagnosis a priori. The general formulas for this may be written:*

$$\frac{100}{1 + \left(\frac{100 - \%\text{ probability of prior disease}}{\%\text{ probability of prior disease}}\right)\left(\frac{\%\text{ probability of antigen positivity without disease}}{\%\text{ probability of antigen positivity with disease}}\right)}$$

and

$$\frac{100}{1 + \left(\frac{100 - \%\text{ probability of prior disease}}{\%\text{ probability of prior disease}}\right)\left(\frac{100 - \%\text{ probability of antigen positivity without disease}}{100 - \%\text{ probability of antigen positivity with disease}}\right)}$$

For example, if there is a 50-50 chance that a person has AS before the B27 test is performed, the positivity of the B27 test raises that likelihood to 92%, whereas negativity of the B27 test reduces the likelihood to 10%. A table of likelihoods for AS with and without B27 positivity is shown below for Caucasians in whom the B27 test has a 10% false negativity and 8% false positivity.

A priori probability of AS (%)	A posteriori probability of AS (%)	
	B27 +	B27−
20	73.8	2.6
50	91.8	9.8
80	97.8	30.3

* These formulations are provided through the courtesy of Professor J. N. Berrettoni.

For Blacks, in whom the B27 test has about a 50% false negativity and 3% false positivity, the probabilities would be as follows:

A priori probability of AS (%)	A posteriori probability of AS (%)	
	B27 +	B27−
20	80.6	11.4
50	94.3	34.0
80	98.5	67.3

It is apparent that B27 positivity in Caucasians and Blacks enhances the diagnosis of AS considerably and almost equally. But B27 negativity in Blacks leaves the diagnosis still very much in doubt, whereas in Caucasians, it strongly reduces the probability of AS.

However, it cannot be overemphasized that the presence of B27 does not establish the diagnosis of AS or any other disease. Careful clinical and radiographic studies must be done in each case. The presence of B27 does not transform into AS back pain caused by trauma, metastatic disease, or emanating from sites in the genitourinary, gastrointestinal, vascular system, and so on. Although the risk for a person with B27 to get AS is about 175 times that of a person without B27, viewed in another way, 80% or more of people with B27 never get AS.

For a physician evaluating a patient who has urethritis, atypical arthritis, no conjunctivitis, and an allergy to penicillin, it must be useful for him to know that his patient has B27 and therefore is more prone to be a candidate for Reiter's disease than for gonococcal urethritis. Nevertheless, this clinical situation must include appropriate smears and cultures for the gonococcus.

In a number of malignancies, it has been suggested that certain HLA antigens appear to have prognostic value. For example, in bronchogenic carcinoma, the presence of Aw19 and B5 appeared to offer a better chance for 2-year survival,[1] whereas in Hodgkin's disease, the same two antigens, Aw19 and B5, occurred more frequently than expected in patients whose initial therapy failed to achieve complete remission and who died within 3 years after diagnosis.[2] In acute lymphocytic leukemia (ALL), A2, which was singled out of retrospective studies, may also have a similar association with better survival. In a prospective study, Cw3 was significantly lower in patients with ALL, thereby suggesting a protective effect. The decrease of both A2 and B12 in multiple sclerosis and of B7 in celiac disease and diabetes mellitus also indicated a degree of resistance. A more absolute resistance to leukemia related to B cell Group 2 reported by Terasaki has not yet been substantiated. This point is discussed further in Malignancy (Chapter 7, Section IX).

A future diagnostic use of HLA typing may be in preventing industrial toxicities and drug hypersensitivity reactions. Possibly no HLA association would be found for the more common effects of these agents (asbestosis or penicillin skin rash) that may have many other contributing factors, but only for the most extreme effects (mesotheliomas after asbestos exposure or anaphylactic reactions to penicillin). Other examples may be those individuals who developed hepatic angiosarcomas following exposure to polyvinyl chloride,[4] those women who developed hepatic tumors after using oral contraceptives,[5] and those who succumbed to Philadelphia's Legionnaires' disease or developed the Guillain-Barré syndrome after a swine flu immunization. Each group might have had similar HLA, Ir, or Ia antigens. Virtually none of this has been explored, but the whole area of industrial toxicity and drug susceptibilities might be very fruitful when viewed in terms of the most severe forms of the disease.

When effective prevention, appropriate therapy, or the elimination of further diagnostic tests can be accomplished, no one would dispute the advantages of early diagnosis of a disease. These advantages are relatively small for the diseases whose diagnosis is enhanced by HLA typing. For example, in AS, attempts can be made to prevent spinal curvature, employ appropriate analgesics, and avoid unnecessary tests. However, very recently, Calin has shown the importance of a simple but carefully designed clinical history as a screening test for ankylosing spondylitis.[6] The preliminary evaluation of their clinical questionnaire showed a 95% sensitivity and 85% specificity, which made it a cheaper and possibly better screening test than B27 testing.[6] The diagnosis of other diseases may also be enhanced either by current HLA testing or some future questionnaire. For example, in Behçet's syndrome, steroids may be used before the disease has advanced; in hemochromatosis, phlebotomy and chelators may be used before excessive iron deposition has damaged other organs; in celiac disease and diabetes mellitus, the disease may be detected early enough to diminish some complications.

The prevention of diseases by HLA testing has obvious eugenic implications that are extremely delicate. For families with diseases linked to HLA, such as C2 and C4 deficiencies, some families with Paget's disease, spinocerebellar ataxia, leprosy, multiple sclerosis, and insulin-dependent diabetes mellitus, genetic counseling would be indicated. It is important to note that although HLA may be linked to one of these diseases in some families, it may not be linked in others because the disease in question may actually have different causes or different inheritance patterns. Consequently, linkage to HLA should be evaluated in each family. For a disease like AS which is not linked to HLA but has a very strong association with the HLA antigen B27, one can estimate the risks for AS in sons of a father with B27 and AS. Before HLA typing was available, the risk for such a son was estimated to be about 15%. With B27, the son's risk for AS is approximately 30%, whereas without it his risk is about 3%.[7] In families with hemochromatosis, unless HLA linkage is found, the present associations with A3 and B14 may help to focus on children with a higher risk for the disease and the need for earlier definitive studies such as a liver biopsy. Bluestone has pointed out that the triad of a negative test for rheumatoid factor, a positive test for B27, and a chronic rheumatic disease are strong indications of a familial seronegative spondyloarthropathy.[8] Finally, some investigators have discussed eliminating sperm possibly having an HLA haplotype that marked the presence of a disease-related antigen. But in view of recent data that show a requirement for two susceptibility genes, this method would have applicability only when both genes are in the *cis* position,[9] and then, only if sperm are proved to have just a single haplotype.[10]

As a technical point in HLA testing, it is important to use more than one antiserum for the determination of any HLA or Ia specificity, and the antisera should be operationally monospecific. A full phenotyping, and ideally genotyping, should be performed. It should be recognized that disease associations with present HLA antigens are generally weak, the linkages are few, and the likelihood is great that two or more genes are involved. The uncertain influence of race, homozygosity, and as yet unnamed HLA and Ia antigens all indicate the vast amount of knowledge still to be acquired in this area through more extensive typing of these individuals. Clearly, the major value of HLA and B-cell typing in disease states is currently in research, and not clinical, areas.

REFERENCES

1. **Rogentine, G.N., Jr., Dellon, A.L., and Chretien, P.B.,** Prolonged disease-free survival in bronchogenic carcinoma associated with HLA Aw19 and B5. A follow-up, in *HLA and Disease,* INSERM, Paris, 1976, 234.
2. **Falk, J. and Osoba, D.,** The HLA system and survival in malignant disease: Hodgkin's disease and carcinoma of the breast, in *HLA and Malignancy,* Murphy, G. P., Ed., Alan R. Liss, New York, 1977, 205.
3. **Johnson, A.H., Ward, F.E., Amos, D.B., Leikin, S., and Rogentine, N.,** HLA and acute lymphocytic leukemia, in *HLA and Disease,* INSERM, Paris, 1976, 227.
4. How hazardous to health is vinyl chloride?, in Medical News, *JAMA,* 228, 1355, 1974.
5. **Mays, E.T., Christopherson, W.M., Mahr, M.M., and Williams, H.C.,** Hepatic changes in young women ingesting contraceptive steroids, *JAMA,* 235, 730, 1976.
6. **Calin, A., Porta, J., Fries, J.F., and Schurman, D.J.,** Clinical history as a screening test for ankylosing spondylitis, *JAMA,* 237, 2613, 1977.
7. **Schaller, J.G. and Omenn, G.S.,** The histocompatibility system and human disease, *J. Pediatr.,* 88, 913, 1976.
8. **Bluestone, R.,** HL-A antigens in clinical medicine, *Dis. Mon.,* 23, 1, 1976.
9. **Dorf, M.E., Maurer, P. H., Merryman, C.F., and Benacerraf, B.,** Inclusion group systems and *cis-trans* effects in responses controlled by the two complementing *Ir-GLΦ* genes, *J. Exp. Med.,* 143, 889, 1976.
10. **Boettcher, B.,** Haploid expression of HLA genes on spermatozoa, *Lancet,* 1, 363, 1977.

Chapter 6

HLA DISEASE ASSOCIATION IN THE PERSPECTIVE OF ABO DISEASE ASSOCIATIONS

The milestone paper by Aird[1] clearly indicated that the frequency of blood group A was greater than the frequency of blood group O in patients suffering from cancer of the stomach when compared to the general population of the same areas. However, despite the very significant differences found ($P < 0.0001$), the relative risk of a person with blood group A getting cancer of the stomach was only 1.26. In individuals with blood group A, there were also increases in alpha hemolytic streptococcal infection, thrombotic diseases, and diabetes mellitus. Similarly, in patients with blood group O, there were small but definite increases in the risk of duodenal ulcers, hemorrhage from duodenal ulcers, hemorrhagic diseases in general, rheumatoid arthritis, and juvenile diabetes mellitus. Despite an enormous amount of work and statistical significance, all of the ABO disease associations have been of minor importance and diagnostically useless.[2,3]

One of the more recent efforts along these lines which has shown perhaps the most significant relative risk (RR) for ABO groups is that for thromboembolism in women taking oral contraceptives.[4] The RR rises quite significantly throughout the world in the A blood group yielding risks of 2.67 in Sweden, 2.76 in the United Kingdom, and 4.59 in the U.S. Lesser but similar trends hold for thromboembolism in pregnant or puerperal women, for coronary thrombosis and myocardial infarction, and for thromboembolism in general. However, the peak RR in a combined analysis for any of these diseases is only 3.12 as reported for thromboembolism in women taking oral contraceptives.

The range of relative risk according to ABO blood groups is at the lower limits of what has been discovered in studies of HLA antigens. In fact, even the peak association of ABO and disease is near the lower level of associations for HLA, many of which exceed this range of relative risk. That disease associations with HLA antigens are so much stronger than those with ABO is no doubt based on the probable analogy with immune response and possibly disease-susceptibility loci being located within or nearby the major histocompatibility complex of man, similar to what has been found in inbred strains of animals.[5]

It should be noted that in special situations for an infection like malaria that has an erythrocytic stage, very potent red-cell resistance factors have been demonstrated for *Plasmodium falciparum* in the form of S hemoglobin and Duffy blood-group negative Fy for *P. vivax*.[6]

REFERENCES

1. **Aird, I., Bentall, H.H., and Fraser Roberts, J.,** The relationship between cancer of the stomach and the ABO blood groups, *Br. Med. J.*, 1, 799, 1953.
2. **Fraser Roberts, J.A.,** ABO blood groups and susceptibility to disease: A review, *Br. J. Prev. Soc. Med.*, 11, 107, 1957.
3. **Wiener, A.S.,** Blood groups and disease, *Am. J. Hum. Genet.*, 22, 476, 1970.
4. **Mourant, A.E., Kopec, A.C., and Domaniewska-Sobczak, K.,** Blood-groups and blood-clotting, *Lancet*, 1, 223, 1971.
5. **Shreffler, D.C. and David, C.S.,** The H-2 major histocompatibility complex and the I immune response region: Genetic variation, function and organization, *Adv. Immunol.*, 20, 125, 1975.
6. **Miller, L.H., Mason, S.J., Clyde, D.F., and McGinniss, M.H.,** Resistance factor to Plasmodium vivax: Duffy genotype Fy Fy, *N. Engl. J. Med.*, 295, 302, 1976.

Chapter 7

CURRENT STATUS OF HLA AND DISEASE ASSOCIATIONS

I. ALLERGY

The inherited propensity to make specific IgE antibody after inhalation or ingestion of minute amounts of allergens is known as atopy. Factors that are important in its development are

1. Genetic factors that include Ir genes and atopy genes
2. The autonomic nervous system
3. Environmental factors
4. Infectious agents
5. A relative immunodeficiency of IgA
6. An abnormality of the mucous membrane

The genetic control of IgE (reagin) production has two bases, an Ir and an atopy gene. The Ir gene control is antigen specific, demonstrates a dose effect of the antigen, involves both IgE and IgG production, and appears to be linked to the major histocompatibility complex or specifically HLA.[1] On the other hand, the atopy gene (or genes) is not antigen specific, does not show a dose effect, affects only IgE production, and is not linked to the major histocompatibility complex.[2]

A. Ragweed Pollinosis (Hay Fever)

In successive generations of seven families, clinical ragweed pollinosis (hay fever) and IgE antibody production specific for antigen E correlated closely with HLA haplotypes.[3] Of 26 family members having the hay fever-associated haplotypes, 20 (77%) had both clinical ragweed hay fever and an intense skin reactivity to antigen E, which is the major antigen found in ragweed pollen. None of the 11 family members who lacked the hay fever-associated haplotype had clinical hay fever, and only 1 had a weak skin reaction to antigen E. The authors concluded that the trait controlled by the suspected Ir gene was inherited as a Mendelian dominant but required additional factors, both genetic and environmental, for full expression. In this study, the proposed linkage of the "Ir-antigen E" gene to HLA was not documented by formal linkage analysis and occurred with seven different HLA haplotypes, only one of which was A1-B8.

In a study of 57 members of a single family of whom 13 had asthma, rhinitis, or both during the ragweed season, there was a significant association between the haplotype A2-B12 and antigen E skin sensitivity as well as possible linkage between the HLA-B locus and the IrE locus.[4] In a study of a 17-member family, 7 had an atopic history: 6 of these had the haplotype A11-B27, and 1 nonatopic had A11 and B27 on different haplotypes, suggesting linkage rather than antigen association.[5] However, dactyl allergy (pollen hypersensitivity) did not segregate with HLA haplotypes in 10 of 12 families, thus providing no support for linkage of this characteristic.[6]

A study by Yoo in ten families with ragweed pollinosis showed that there was a strong association within families between ragweed pollinosis and an HLA haplotype, though no significant relationship to any one particular haplotype was noted.[7] Ragweed pollinosis was found with the same HLA haplotype within a family on 27 of 44 occasions but in only 1 of 24 individuals without the hay fever-associated haplotype. Even though the susceptibility haplotype did not occur in every patient who was symptomatic, the association was highly significant ($P<0.005$). No formal linkage analysis was performed. There was no correlation with any specific HLA antigen or specific HLA haplotype among families. In this study, PHA stimulation was performed on lymphocytes from 102 asymptomatic individuals and 25 symptomatic individuals within these families. A significantly greater response to PHA occurred in those individuals who had pollinosis (P<0.025). However, there was no association between the PHA results and the inheritance of any HLA antigen or haplotype. In 105 patients examined for skin sensitivity to ragweed antigen 5 (Ra5), 18 were highly sensitive to Ra5, 77 were Ra5 insensitive, and 10 patients had an intermediate type of reaction.[8] Comparing the Ra5-sensitive group with the Ra5-insensitive group, there was found to be a significant increase of B7 with a frequency of 50% (9/18) in the former group compared to 19.5% (15/77) in the latter group (P<0.02). When one examined four cross-reacting antigens of the B7 group, namely, B7, plus Bw22, B27, and B40, the differences were even more significant (P< 0.006). On the other hand, individuals with A10, A9, and B13 tended to show a degree of negative association with Ra5 sensitivity, although none of these relationships were significant. These authors interpreted their data as showing that an Ir gene (or genes) was strongly related to the B7 cross-reacting group of HLA antigens.

Thus, there is substantial evidence that skin sensitivity to two antigens derived from ragweed pollen, namely, antigen E and Ra5, are related to the HLA system, possibly by linkage in the former[3,4] and by association in the latter.[8]

A further study of 30 Ra5-reactive and 84 Ra5-nonreactive individuals showed increased frequencies of B7 and the B7 cross-reacting group (50 and 70%, respectively), compared to 18 and 37% for the nonreactors (P<0.001 and 0.003, respectively).[9] These same authors have also studied patients in another area (Bellville, Illinois) and found that there was still an increased frequency of B7 and the B7 cross-reacting group (31 and 63%, respectively) in 35 Ra5 reactors compared to 15 and 29%, respectively, for the 41 Ra5 nonreactors.[9]

Another purified ragweed antigen, Ra3, and Rye 1 were also studied.[10] When compared to 64 patients insensitive to Rye 1, 136 Caucasian patients sensitive to Rye 1 grass pollen antigen showed a significant elevation in A1 (P< 0.03), B5 (P<0.0005), and the A1-B8 phenotype (P<0.007). In 76 Caucasians reactive to Ra3, when compared to 40 Ra3-nonreactive individuals, it was found that A2 (P< 0.04), B12 (P<0.07), and the A2-B12 phenotype (P<0.06) were significantly elevated in the Ra3-reactive group. When subgroups of these allergic patients were examined, namely, those who had a genetically low total serum IgE level, it was found that 14 of 15 Ra3-reactive compared to 4 of 15 Ra3-nonreactive subjects had A2, and the only Ra3-reactive subject without A2 had B12. Thus, it appeared that IgE antibody responses to Rye 1 and Ra3 in people with limited total IgE levels were associated with two common Caucasian phenotypes, A1-B8 and A2-B12, respectively, and conversely, a low total IgE level appeared to be a necessary substrate for the demonstration that an HLA antigen was significantly associated with responsiveness to Rye 1 or Ra3. Within each relevant haplotype, there appeared to be a difference in the series of the antigen associated with each reactivity in that the Rye 1 sensitivity was more closely related to the B allele B8, whereas the Ra3 sensitivity appeared to be more closely related to the A allele A2. These authors postulated that there were two Ir loci, one controlling the response to Rye 1 that was closer to the B locus and the other controlling the response to Ra3 located closer to the A locus.

B. Asthma

One of the earliest studies of allergic diseases was that by Thorsby in which 35 patients with childhood asthma showed an increase in the haplotype A1-B8.[11] A1 occurred with a frequency of 42.9% in the 35 asthmatics compared to 28.3% in 891 controls, and B8 occurred with a frequency of 31.4% compared to 26.1% of controls. There was also an increased frequency of an antigen detected by an antiserum named FJH-AJ, which was shorter than B27 and occurred in 11.4% of the asthmatics compared to 4.2% of controls. In this group of 35, 4 of the 6 patients who had the more serious endogenous form of asthma, in which no specific exogenous allergen could be incriminated, possessed the A1-B8 haplotype.

Later studies by Morris of 47 patients diagnosed as having bronchial asthma based on the presence of variable airway obstruction, response to inhaled beta sympathomimetic bronchodilators and steroids, and on blood or sputum eosinophilia showed that there were significant differences between the HLA antigen frequency in two subgroups of asthma: extrinsic (21 patients) and intrinsic (26 patients).[12] Compared to control values for the five antigens A1, B8, B12, B40, and B15 of 29, 25, 23, 10, and 9%, respectively, patients with intrinsic asthma had the following frequencies of these antigens: 29, 29, 19, 29, and 38%; patients with extrinsic asthma had the following frequencies of these same antigens: 46, 42, 50, 4, and 15%. Therefore, the major differences were an increase in A1, B8, and B12 in extrinsic asthma and an increase in B40 and B15 in intrinsic asthma.

Rachelefsky found that in 71 asthmatics, B8 occurred in only 7% compared to 21% of controls, but he did not distinguish any subgroups.[13] These same authors used B-cell antisera to study a group of 30 families (151 individuals), 24 of which had one allergic parent and all of which had one child with extrinsic asthma. All five of the known Terasaki B-cell specificities were found in a higher proportion of asthmatic patients than in normals, but in particular, the B group 2 was most strikingly elevated (88% of asthmatic children compared to 24% of controls — $P<0.005$ with correction). These authors concluded that all B-lymphocyte specificities, or at least those thus far described, were increased in asthmatic patients, suggesting that they may be more readily exposed on the surface of lymphocytes from asthmatics compared to controls. A linkage of asthma to the B lymphocyte group 2 antigen was not convincingly shown in an analysis of seven of the nine families examined, and recent evidence indicates that the validity of B group 2 is uncertain.

Recently, HLA-Bw6, an antigen believed to be in linkage disequilibrium with B-locus antigens, has been found in homozygous form in 81% (21 of 26) of patients with intrinsic asthma. This finding suggested that intrinsic asthma may be a recessive disease.[14]

C. Hypersensitivity Pneumonitis

Hypersensitivity pneumonitis, which develops only in a minority (less than 5%) of individuals exposed to antigens in inhaled organic dust, showed a remarkable familial incidence in a study of 20 patients.[15] Also interesting was the fact that specific T-cell response to avian antigen measured by in vitro blastogenesis was inhibited by HLA antisera directed against B- rather than A-locus determinants. B40 occurred in 35% of the 20 unrelated patients compared with 10% of 735 normal controls ($P<0.005$) with a relative risk of approximately 4.8. In two families, the disease might have been increased in those with HLA haplotypes A2-B15, A2-B40, and A1-B8, and protected against in those with A3-B7.[16]

D. Penicillin Allergy

In a study performed on 38 patients with known penicillin allergy, no significant association was found with any HLA antigen of the A or B series.[17]

E. Tuberculin Hyporesponsiveness in BCG Treatment

In patients with malignant melanoma, the average area of induration to 5 TU of

PPD at 48 hr in B7-positive patients was about one half the area of induration that developed in non-B7 patients ($P<0.018$) before BCG treatment as well as after treatment when the difference was even more pronounced ($P<0.0007$.)[18] However, post-BCG patient survival differences between B7-positive and B7-negative individuals were not significant. These data suggested that B7 patients have a specific hyporesponsiveness in delayed cutaneous sensitivity to PPD tuberculin. In addition, B7-positive patients have cutaneous hyporesponsiveness to many microbiologic skin test allergens after BCG immunotherapy.

It should be noted that a B7-associated tuberculin unresponsiveness has also been described in patients with sarcoidosis (see Pulmonary [Chapter 7, Section XII]).[19]

The general decrease in responsiveness to PPD in those with B7 even without known disease might be useful information in diagnosing tuberculosis, a disease that strangely enough has not yet been tested for HLA association.

F. Milk Intolerance

A single study of 21 patients showed no significant alteration in HLA antigen frequency.[20]

G. Atopic Dermatitis (see Dermatology [Chapter 7, Section V])

REFERENCES

1. **Levine, B.B.,** *Biochemistry of Acute Allergic Reaction,* Austin, K.F. and Becker, E.L., Eds., Blackwell, Oxford, 1971.
2. **DeWeck, A.L.,** Summary of the Allergy Workship of the HLA and Disease Symposium, Paris, June 1976.
3. **Levine, B.B., Stember, R.H., and Fotino, M.,** Ragweed hay fever: genetic control and linkage to HL-A haplotypes, *Science,* 178, 1201, 1972.
4. **Blumenthal, M.N., Amos, D.B., Noreen, H., Mendell, N.R., and Yunis, E.J.,** Genetic mapping of Ir locus in man: linkage to second locus of HL-A, *Science,* 184, 1301, 1974.
5. **Geerts, S.J., Pöttgens, H., Limburg, M., and van Rood, J.J.,** Predisposition for atopy or allergy linked to HL-A, *Lancet,* 1, 461, 1975.
6. **Pillier-Loriette, C., Marcelli-Barge, A., Dausset, J., Treich, I., Gervais, P., Raffard, M., Henocq, E., Berman, D., and de Montis, G.,** Search for a correlation between familial allergy to dactyl (pollen hypersensitivity) and HLA antigens, *Tissue Antigens,* 8, 87, 1976.
7. **Yoo, T.J., Flink, R.J., and Thompson, J.S.,** The relationship between HL-A antigens and lymphocyte response in ragweed allergy, *J. Allergy Clin. Immunol.,* 57, 25, 1976.
8. **Marsh, D.G., Bias, W.B., Hsu, S.H., and Goodfriend, L.,** Association of the HL-A7 cross-reacting group with a specific reaginic antibody response in allergic man, *Science,* 179, 691, 1973.
9. **Goodfriend, L., Santilli, J., Jr., Schacter, B., Bias, W.B., and Marsh, D.G.,** HLA-B7 cross-reacting group and human IgE mediated sensitivity to ragweed allergen Ra5, in *HLA and Disease,* INSERM, Paris, 1976, 178.
10. **Marsh, D.G., Chase, G.A., and Bias, W.B.,** "Mapping" of postulated Ir genes within HLA by studies in allergic populations, in *HLA and Disease,* INSERM, Paris, 1976, 181.
11. **Thorsby, E., Engeset, A., and Lie, S.O.,** HL-A antigens and susceptibility in diseases. A study of patients with acute lymphoblastic leukaemia, Hodgkin's disease and childhood asthma, *Tissue Antigens,* 1, 147, 1971.
12. **Morris, M.J., Vaughan, H., Lane, D., and Morris, P.J.,** HLA and asthma, in *HLA and Disease,* INSERM, Paris, 1976, 182.
13. **Rachelefsky, G., Park, M.S., Siegel, S., Terasaki, P.I., Katz, R., and Saito, S.,** Strong association between B-lymphocyte group-2 specificity and asthma, *Lancet,* 2, 1042, 1976.

14. **Brostoff, J., Mowbray, J.F., Kapoor, A., Hollowell, S.J., Rudolf, M., and Saunders, K.B.,** 80% of patients with intrinsic asthma are homozygous for HLA W6, *Lancet*, 2, 872, 1976.
15. **Allen, D.H., Basten, A., Woolcock, A.J., and Guinan, J.,** HLA and bird breeder's hypersensitivity pneumonitis, in *HLA and Disease*, INSERM, Paris, 1976, 173.
16. **Allen, D.H., Basten, A., Williams, G.V., and Woolcock, A.J.,** Familial hypersensitivity pneumonitis, *Am. J. Med.*, 59, 505, 1975.
17. **DeWeck, A.L. and Spengler, H.,** Evaluation of genetic control on the immune response to penicillin in man, in *HLA and Disease*, INSERM, Paris, 1976, 177.
18. **Buckley, C.E., III, White, D.H., and Siegler, H.F.,** HL-A B7 associated tuberculin hyporesponsiveness in BCG treated patients, in *HLA and Disease*, INSERM, Paris, 1976, 175.
19. **Persson, I., Ryder, L.P., Staub Nielsen, L., and Svejgaard, A.,** The HL-A7 histocompatibility antigen in sarcoidosis in relation to tuberculin sensitivity, *Tissue Antigens*, 6, 50, 1975.
20. **Dausset, J. and Hors, J.,** Some contributions of the HL-A complex to the genetics of human diseases, *Transplant. Rev.*, 22, 44, 1975.

II. CARDIOVASCULAR DISEASES

Most of the studies in this area have investigated heterogeneous clinical diseases and no confirmation of these results is yet available.

A. Essential Hypertension

In 107 patients with essential hypertension, B12 was increased to 41% compared to 28% of controls, a difference that was not significant when corrected for the number of antigens tested. However, hypertensives were not identified or excluded from the control population which, therefore, was not an appropriate control.[1]

Another study of 144 patients from Australia showed increases primarily in B18, Bw21, and B13 with relative risks of 4.0, 3.2, and 2.1, respectively.[2]

A third study found that B15 was increased twofold in patients with a family history of hypertension (corrected $P<0.05$).[3]

B. Coronary Artery Disease

One hundred men ranging in age from 39 to 80 years with a history of a myocardial infarction showed no significant alteration in any HLA antigen frequency, though B27 was slightly increased.[4]

In 198 Australian patients with some manifestation of coronary artery disease, there were increases in Bw21, B18, and B13 with relative risks of 5.2, 2.4, and 2.0, respectively.[2]

Different results were reported by Mathews, who found that death rates from ischemic heart disease paralleled the frequency of B8 and the haplotype A1-B8 in at least ten populations.[5] The level of cholesterol in Finns appeared related to B8 and B15. The relationship of B8 and B15 to such a lipid abnormality raises the question of whether the involvement of these same antigens with juvenile diabetes mellitus is because of an associated metabolic lipid disorder.[5] However, 100 patients with type IIa hyperlipoproteinemia and hypercholesterdemia showed increased frequencies of B17 and Bw35.[6]

C. Congenital Heart Disease

A study by Buc showed an increase in A2 in patients with a variety of congenital heart diseases.[7] The frequency of A2 in this congenital heart study was 75% compared to 44% in controls, a significant difference (corrected $P<0.001$). There was not a selective A2 association among the various types of congenital heart disease.

A preliminary study of three families with secundum-type atrial septal defects suggested linkage of a cardiac development gene with the HLA loci.[8]

D. Rheumatic Heart Disease

Gorodezky's study of rheumatic heart disease secondary to documented rheumatic fever suggested a significant decrease in the A3 antigen, since none of their 48 patients had this antigen.[9] An early study of rheumatic fever and rheumatic heart disease also disclosed a decreased frequency of A3, an increased number of shared parental antigens, and consequently, an increased frequency of homozygosity in patients.[10] But a later study could not confirm any of these findings and reported that there was a decrease in A28 and an increase in B17 in European patients, and that there was a decrease in A10 and an increase in A3 and B8 in Maori patients.[11] Ward's study of 133 patients with acquired valvular heart disease showed no decrease in A3, but in those with no history of rheumatic fever, there were increases in A29 and Aw30/31, cross-reacting antigens of the Aw19 group.[12] Thus, in rheumatic heart disease, there is only a suggestion of A3 being diminished in frequency and numerous conflicting data (see Infections and Immunizations [Chapter 7, Section VIII]).

E. Arterial Occlusive Disease

In 155 Japanese patients, three types of arterial occlusive disease were studied: thromboangiitis obliterans, arteriosclerosis obliterans, and chronic arterial occlusive disease.[13] In 28 patients with thromboangiitis obliterans, there were significant increases in A9 from 42.5% to 75%, in B40 from 18.8% to 42.9%, and in the Japanese split of Bw22, known as BJW 22-2 (J-1), from 18.2% in controls to 46.4% in patients. In the same patients, A12 was significantly decreased to 0% from 19.9% in controls. In patients with arteriosclerosis obliterans,[14] the BJW 22-1 split of 22 was insignificantly increased and Cw1 was significantly lower (8.5% compared to 36.3% in controls, corrected $P<0.01$). Finally, in patients with chronic arterial occlusion, no significant HLA antigen differences were noted.

F. Takayasu's Disease

A Japanese study of Takayasu's disease (pulseless disease) that is of unknown etiology and typically found in young females mainly from Asian countries showed that in five families, in each of which two members suffered from Takayasu's disease, there was a common haplotype containing either B5 or B40.[15] A set of identical twins, both with Takayasu's disease, contained one HLA haplotype that did not exist in other healthy sisters in the family. A further study of 24 unrelated patients showed an increased frequency of B5 (59% compared to 33% in healthy controls) and B40 (53% compared to 34%).

G. Buerger's Disease

In 90 Japanese patients with Buerger's disease, B5 tended to be increased but not significantly.[16]

H. Temporal Arteritis

In 61 unrelated patients with temporal arteritis (Horton's disease), Seignalet found that B14 was increased to 22.95% compared to 8.67% in controls (corrected $P<0.05$),[17] but B14 was found in only 4% of 50 patients with polymyalgia rheumatica who did not have temporal arteritis. Conversely, Bw38 and B5 were increased in polymyalgia but not in temporal arteritis alone. These two different antigen patterns in isolated temporal arteritis and in polymyalgia rheumatica suggested that these two entities were distinct despite their frequent clinical concurrence.

I. Idiopathic Aortic Insufficiency

Idiopathic aortic insufficiency, despite its relationship to ankylosing spondylitis, failed to show the suspected high frequency of B27 or any other significant antigen frequency deviation.[18]

J. Mitral Valve Prolapse

Thirty-one unrelated Caucasian patients with mitral valve prolapse (MVP) diagnosed by a systolic click and/or murmur plus echocardiographic evidence of MVP were tested for the frequencies of 28 HLA antigens.[19] Other features of MVP may include chest pain, arrhythmias, nonspecific ST-T wave changes, skeletal abnormalities, mitral regurgitation, and myxomatous changes in the mitral valve.

Two HLA antigens were found to occur with increased frequency: Bw35 (45.2% compared to 19.5% of controls [uncorrected $P<0.01$, corrected NS]) and A3 (35.5% compared to 26.7% of controls [NS]).[19] However, the most interesting finding was that seven patients (22.6%) had an A3-Bw35 phenotype compared to 6.3% of controls (uncorrected $P < 0.01$).

The relative risk (RR) for this disease conferred by A3 was 1.52; by Bw35, 3.38; and by the A3-Bw35 phenotype, 4.27.[19] It is possible that the A3-Bw35 phenotype might identify a subgroup of MVP patients with: (1) a common etiology, (2) other organ involvement, (3) a different prognosis, or (4) familial occurrence.

REFERENCES

1. **Gelsthorpe, K., Doughty, R.W., Bing, R.F., O'Malley, B.C., Smith, A.J., and Talbot, S.,** HL-A antigens in essential hypertension, *Lancet,* 1, 1039, 1975.
2. **Mathews, J.D., England, J., Shaw, J., Hunt, D., Mathieson, I.D., Cowling, D.C., and Tait, B.D.,** Antigen and haplotype frequencies in essential hypertension and ischaemic heart diesase, in *HLA and Disease,* INSERM, Paris, 1976, 257.
3. **Kristensen, B., Andersen, P. L., Lamm, L. U., and Kissmeyer-Nielsen, F.,** HLA antigens in essential hypertension, *Tissue Antigens,* 10, 70, 1977.
4. **Scott, B.B., McGuffin, P., Rajah, S.M., Stoker, J.B., and Losowsky, M.S.,** Histocompatibility antigens and myocardial infarction, *Tissue Antigens,* 7, 187, 1976.
5. **Mathews, J.D.,** Ischaemic heart-disease: Possible genetic markers, *Lancet,* 2, 681, 1975.
6. **Raffoux, C., Pointel, J. P., Drovin, P., Streiff, F., Debry, G., and Sauvanet, J. P.,** Type IIa hyperlipoproteinemia and the HLA system, *Tissue Antigens,* 11, 55, 1978.
7. **Buc, M., Nyulassy, S., Stefanovic, J., Jakubcova, I., and Benedekova, M.,** HL-A2 and congenital heart malformations, *Tissue Antigens,* 5, 128, 1975.
8. **Mohl, W. and Mayr, W. R.,** Atrial septal defect of the secundum type and HLA, *Tissue Antigens,* 10, 121, 1977.
9. **Gorodezky, C.,** HLA and rheumatic heart disease, in *HLA and Disease,* INSERM, Paris, 1976, 34.
10. **Falk, J.A., Fleischman, J.L., Zabriskie, J.B., and Falk, R.E.,** A study of HLA antigen phenotype in rheumatic fever and rheumatic heart disease patients, *Tissue Antigens,* 3, 173, 1973.
11. **Caughey, D.E., Douglas, R., Wilson, W., and Hassall, I.B.,** HL-A antigens in Europeans and Maoris with rheumatic fever and rheumatic heart disease, *J. Rheumatol.,* 2, 319, 1975.
12. **Ward, C., Gelsthorpe, K., Doughty, R.W., and Hardisty, C.A.,** HLA antigens and acquired valvular heart disease, *Tissue Antigens,* 7, 227, 1976.
13. **Juji, T., Ohtawa, T., Kawano, N., Mishima, Y., Tohyama, H., and Ishikawa, K.,** HLA antigens in Japanese patients of arterial occlusive diseases, in *HLA and Disease,* INSERM, Paris, 1976, 251.
14. **Ohtawa, T., Juji, T., Kawano, N., Mishima, Y., Tohyama, H., and Ishikawa, K.,** HL-A antigens in thromboangiitis obliterans, *JAMA,* 230, 1128, 1974.
15. **Numano, F., Isohisa, I., Maezawa, H., and Juji, T.,** Takayasu's disease and HL-A typing, in *HLA and Disease,* INSERM, Paris, 1976, 259.

16. **Hoshino, K., Inouye, H., Unokuchi, T., Ito, M., Tamaoki, N., and Tsuji, K.,** HLA and diseases in Japanese, in *HLA and Disease,* INSERM, Paris, 1976, 249.
17. **Seignalet, J., Janbon, C., Sany, J., Janbon, F., Bidet, J.M., Brunel, M., Jourdan, J., and Bussiere, J.L.,** HLA in temporal arteritis. *Tissue Antigens,* 9, 69, 1977.
18. **Calin, A., Fries, J.F., Stinson, F.B., and Payne, R.,** Normal frequency of HLA B27 in aortic insufficiency, in *HLA and Disease,* INSERM, Paris, 1976, 21.
19. **Braun, W.E., Ronan, J., Schacter, B., Gardin, J., Isner, J., and Grecek, D.,** HLA antigens in mitral valve prolapse, *Transplant. Proc.,* 9, 1869, 1977.

III. COMPLEMENT DEFICIENCIES AND IMMUNODEFICIENCY DISEASE

A. Complement Deficiencies

Numerous reports have now confirmed the fact that deficiencies of certain components of the complement system are related to specific HLA antigens and in some cases with an HLA haplotype. In brief, the data is best for the control of C2, C4, and C8 *levels* by genes in the HLA region and for linkage of the C2 structural gene to HLA. The complement components will be covered numerically.

In one family with C1 esterase inhibitor deficiency (hereditary angioneurotic edema) inherited as an autosomal dominant, the proband and her eldest son with the disease both had an identical haplotype, A2-B5, but in addition, a younger son who had normal C1 esterase inhibitor activity also had the A2-B5 haplotype. [1] Similarly, in a second family, the proband and her eldest son with the disease both had an identical haplotype, Aw19-B40-Cw3, but a normal younger son also had this same haplotype. Together these preliminary results suggest that there is no apparent linkage between C1 esterase inhibitor deficiency and HLA. Similarly, no HLA linkage was found in another family with C1r deficiency.[2]

Perhaps the best data for linkage between HLA and a complement component deficiency exist for C2. In studies that expanded their original findings in families with C2 deficiency, Fu studied three families in whom the proposita were homozygous for C2 deficiency and one family in which the propositus was heterozygous.[3] All four of these individuals had the A10-B18 haplotype, which has a frequency of less than 1% in a normal population. Three C2-deficient homozygotes from three of these families were mutually nonreactive in mixed lymphocyte culture reactions, and the heterozygous C2 deficient individual from the fourth family was nonreactive to stimulation by one of three homozygotes tested. This common MLC determinant was found to be Dw2, then known as LD-7a.

The single largest study of ten families with C2 deficiency, nine of whom were heterozygous for the C2 deficiency, revealed that in eight of the families, all members who were C2 deficient had the A10 antigen; in nine families, all members who were C2 deficient had the B18 antigen; and in six families, the C2 deficiency was associated with the A10-B18 haplotype.[4] The fact that all patients were found to be C2-deficient when possessing the A10-B18 haplotype suggested a high rate of linkage disequilibrium. This association diminished when B18 alone was present and was even less when A10 alone was present, indicating that the B locus of HLA was closer to the C2 locus than was the A locus. This study also reinforced the finding of a high proportion of recombinants in C2-deficient individuals, with one family showing a B-D recombination and the other a D-C2 locus recombination. However, a family has been reported with both homozygous and heterozygous C2-deficient individuals in which the C2 deficiency was inherited with a haplotype A3-B5, Aw30-B13, or both but segregated independently of an A10-B18 haplotype that also existed in the family.[5]

Although the initial patients described with C2 deficiency had no apparent clinical illness, later reports revealed an occurrence of discoid or systemic lupus, polymyositis, and chronic anaphylactoid purpura. Three homozygous C2-deficient individuals had manifestations of systemic lupus erythematosus.[3] A detailed description of a different clinical syndrome associated with C2 deficiency was reported by Friend. [6] This patient, who had mononeuropathy multiplex, arthritis, Raynaud's phenomenon with chronic vasculitis, and a history of recurrent pneumonia and bronchitis, was a presumptive A10-B18 homozygote as was one of the patients reported by Fu.[3]

Both systemic lupus and multiple sclerosis, the latter of which has been shown to have an association with Dw2 also, have some suppression of cell-mediated immunity as part of their pathologic process that possibly may be reflected in the Dw2 gene.[7]

In Fu's study, a possible recombinant who had the haplotype A10-B18 but lacked the Dw2 antigen was not C2-deficient, a finding that would place the C2 gene close to that for the D series of antigens.[3] Another possible recombinant was reported between the D and C2 loci, which supported a high recombination frequency (4%) in the 50 meiotic divisions of these four families.[8]

In a population study of the incidence of heterozygous C2 deficiency by Mowbray, it was found that the defect exists not only with the A10-B18 haplotype but also frequently with B18 and Bw21 alone.[9] Surprising was the fact that 9 of 25 "normal people" with B40 have been found to have approximately half-normal levels not only of C2 but also of C4 and properdin factor B. Because patients with C2 deficiency may have a failure to clear immune complexes from the circulation, they may suffer from immune complex diseases. Evidence for this was presented by Mowbray, who reported that the A10-B18 haplotype as well as B18 and Bw40 were all increased in renal allograft recipients whose original disease was immune complex nephritis. These authors also found a high frequency of atopic disease in otherwise normal people who had a heterozygous C2 deficiency.

Two studies have investigated C2 deficiency occurring in other diseases, namely, multiple sclerosis [7] and systemic and discoid lupus erythematosus.[10] As described under Multiple Sclerosis (Chapter 7, Section X.A), the C2 deficiency in Dw2-positive MS patients showed a bimodal distribution of C2 levels. Approximately one half normal C2 levels were found in one segment of the Dw2-positive MS group and normal C2 concentrations in the other segment, indicating a high percentage of a heterozygous C2 defect in multiple sclerosis which in that disease, was associated with the antigen complex A3-B7-Dw2. On the other hand, Agnello found that C2 deficiency led to a form of systemic lupus which lacked some of the immune responses characteristically seen in patients with classic SLE.[10]

In summary, the vast majority of information indicates a very strong association of C2 levels with the A10-B18 haplotype, B18 itself, and even A10. Furthermore, the C2 locus and HLA appeared to be linked. A single study showing the association of Dw2 with the disease in the proband of each of four families is extremely interesting but needs confirmation, as does the indication that the recombination frequency is high.

A study of C4 deficiency in a 5-year-old boy with a lupus-like syndrome, nephritis, and antinuclear antibody showed that in three generations of this family, eight individuals were heterozygous for C4 deficiency with half-normal levels of C4.[11] The C4 deficiency gene was associated with A2-B12-Dw2 on the maternal side and A2-B15-LD108 on the paternal side. Another study of C4 deficiency by Rittner found that in one family, the homozygous deficient state in one patient was associated with A2-B40-

Cw3 homozygosity, the heterozygous state in three patients with the A2-B40-Cw3 haplotype, and normal C4 levels in two patients with the absence of that same haplotype.[12]

In C6 deficiency, a Boston study of seven unrelated patients showed that five of the seven had Aw24, whereas no others in the family did.[13] However, a patient with total C6 deficiency had the same A9-B15 as did his sibling with normal C6. The other HLA haplotype of the patient's mother who had half-normal C6 level also was not accompanied by any C6 deficiency. [2] Consequently, the linkage of C6 and HLA is uncertain at present.

In a study of a 14-member family with homozygous and heterozygous deficiency of the seventh component of complement, no association with an HLA haplotype was found. [14] Another unrelated patient has also been reported to have a complete C7 deficiency unassociated with HLA. [15]

C8 deficiency in a patient who had xeroderma pigmentosum and was from a 23-member Tunisian family showed that there was no association between the C8-defective chromosome and any HLA antigen.[16] D-locus studies in this family revealed that the homozygous C8-deficient patients were not D homozygotes. A peculiar twist in this family was the fact that the three C8 homozygous-deficient individuals were clinically well, whereas the two C8 heterozygote individuals had xeroderma pigmentosum. These results conflict with those of another three-generation family in which the C8 deficiency was associated with HLA haplotypes, A2-Bw35 and Aw33-Bw35, inherited from the homozygous deficient mother. [17] It has been suggested that the difference between these two studies may rest in the fact that since the C8 molecule has three polypeptide chains, only one of them may be closely linked to the HLA locus [16,17] or that an inhibitor rather than a deficiency is involved.

HLA and properdin factor B (Bf) are linked, and the latter may be involved in some glomerulopathies [18] (see Renal Diseases [Chapter 7, Section XIII]).

B. Immunodeficiency Diseases

In 26 patients with a selective deficiency of IgA, there was a suggestion of an increased frequency of A1 and B8, but no control data were provided. [19] Ataxia-telangiectasia, an autosomal recessive disease, was studied in 30 patients, their parents, and healthy siblings in whom the deviant antigens were B17 (53.3% of patients compared to 23.9% of controls), A1 (3.3% vs. 25.3% controls), and B8 (0% compared to 10.8% of controls).[20]

In studies of collections of primary immunodeficiency diseases, Hors reported an increased frequency of A1 in 36 children with immunodeficiency disease (47% compared to 21% of controls, corrected $P<0.02$) that was similar to the 42% frequency in 19 individual case studies reported by others. [21] In a study by Buckley of 52 patients with various types of immunodeficiencies, including severe combined immunodeficiency, infantile x-linked agammaglobulinemia, common variable agammaglobulinemia, ataxia-telangiectasia, x-linked immunodeficiency with hyper-IgM, Wiskott-Aldrich syndrome, hyper-IgE syndrome, selective IgA deficiency, Nezelof's syndrome, and Candida granuloma, 5 of the 43 families involved had two HLA-identical affected children, 2 families had affected children differing by one haplotype, and 2 other families had two affected children differing by two haplotypes.[22] Furthermore, seven families had HLA-identical normal siblings. Overall there was no particular antigen that was increased in frequency in any of the various categories of immunodeficiency in Buckley's study.[22] They showed only an increased frequency of A2 (36/52 patients; 70% compared to 48% of normals), a decreased frequency of A3 (7/52 patients; 13% compared to 29% in normals), and a decreased frequency of A1 (11/52 patients; 21% compared to 31% of normals). However, of interest was the fact that the observed frequencies of all 12 A-series alleles which were studied in this investigation were significantly different from those in healthy controls, and the major departure in this profile of antigens was contributed by A2 and A3. The finding of two 2-haplotype different pairs of siblings with agammaglobulinemia, one x-linked and one common variable, indicated that genes determining these immune deficiencies are probably not associated with HLA. However, the distortion of the frequencies of A-series antigens generally agrees with the suggestion of Hors that a genetic locus closely associated with the A locus may be involved in the control of immune differentiation.[21] A report of the International Bone Marrow Transplant Registry noted an increase in B7 (39% compared to 18% of controls) in 57 Caucasians with severe combined immunodeficiency disease.[23]

A study from Germany of 13 patients with a variety of immune deficiency diseases also showed an increase in the frequency of A2 to 92.3% of the patients compared to 60.8% of controls, and a decrease in B8 to 0% compared to 8% of controls.[24]

Fifty-eight patients with idiopathic paraproteinemia or essential monoclonal gammopathy were shown to have an increased frequency of B7 (40% compared to 24% in 285 controls) with a relative risk of 2.1.[25] The increased frequency of B7 here and in multiple sclerosis suggests a common theme of decreased cellular immunity and abnormal antibody production, possibly the effect of different populations of suppressor cells. In contrast, B8 frequency appears more closely related to antibody production with low levels of antibody in immune deficiency diseases and high levels of antibody in endocrinologic diseases and HBsAg infection.

One of the earliest reports of HLA typing in immunodeficiency diseases by Terasaki reported an excessive number of A- and B-series antigens.[26] Terasaki now believes that this finding represented the presence of B-cell antibodies in the HLA typing sera which would react with B antigens on the patient's cells and appeared, at that time, like extra HLA reactions.[27]

REFERENCES

1. **Tanimoto, K., Horiuchi, Y., Juji, T., Yamamoto, K., Kodama, J., Murata, S., Funahashi, S., and Nagaki, K.,** HLA types in the two families with hereditary angioneurotic edema, in *HLA and Disease,* INSERM, Paris, 1976, 212.
2. **Mittal, K.K., Wolski, K.P., Lim, D., Gewurz, A., Gewurz, H., and Schmid, F.R.,** Genetic independence between the HL-A system and deficits in the first and sixth components of complement, *Tissue Antigens,* 7, 97, 1976.
3. **Fu, S.M., Stern, R., Kunkel, H.G., Dupont, B., Hansen, J.A., Day, N.K., Good, R.A., Jersild, C., and Fotino, M.,** Mixed lymphocyte culture determinants and C2 deficiency: LD-7a associated with C2 deficiency in four families, *J. Exp. Med.,* 142, 495, 1975.
4. **Gibson, D.J., Glass, D., Carpenter, C.B., and Schur, P.H.,** Hereditary C2 deficiency: Diagnosis and HLA gene complex associations, *J. Immunol.,* 116, 1065, 1976.
5. **Wolski, K.P., Schmid, F.R., and Mittal, K.K.,** Genetic linkage between the HL-A system and a deficit of the second component (C2) of complement, *Science,* 188, 1020, 1975.
6. **Friend, P., Repine, J.E., Kim, Y., Clawson, C.C., and Michael, A.F.,** Deficiency of the second component of complement (C2) with chronic vasculitis, *Ann. Intern. Med.,* 83, 813, 1975.
7. **Bertrams, J., Opferkuch, W., Grosse-Wilde, H., Netzel, B., Rittner, Ch., Kuwert, E., and Schuppien, W.,** HLA linked C2 deficiency in multiple sclerosis (MS), in *HLA and Disease,* INSERM, Paris, 1976, 295.
8. **Day, N.K., l'Esperance, L., Good, R.A., Michael, A.F., Hansen, J.A., Dupont, B., and Jersild, C.,** Hereditary C2 deficiency. Genetic studies and association with the HLA system, *J. Exp. Med.,* 141, 1464, 1975.

9. **Mowbray, J.F.**, Association of heterozygous C2 deficiency with both disease and HLA, in *HLA and Disease*, INSERM, Paris, 1976, 204.
10. **Agnello, V.**, Association of C2 deficiency (C2D) and HLA genes with systemic and discoid lupus erythematosus (SLE, DLE), in *HLA and Disease*, INSERM, Paris, 1976, 296.
11. **Ochs, H.D., Rosenfeld, S.I., Thomas, E.D., Giblett, E.R., Alper, C.A., Dupont, B., Hansen, J.A., Grosse-Wilde, H., and Wedgwood, R.J.**, Linkage between the gene(s) controlling the synthesis of C4 and the major histocompatibility complex, in *HLA and Disease*, INSERM, Paris, 1976, 208.
12. **Rittner, C., Hauptmann, G., Grosse-Wilde, H., Grosshans, E., Tongio, M.M., and Mayer, S.**, Linkage between HL-A (Major Histocompatibility Complex) and genes controlling the synthesis of the fourth component of complement, in *Histocompatibility Testing 1975*, Kissmeyer-Nielsen, F., Ed., Munksgaard, Copenhagen, 1975, 945.
13. **Carpenter, C.B.**, personal communication, 1976.
14. **Delage, J. H., Bergeron, P., Simard, J., Lehner-Netsch, G., and Prochaska, E.**, Hereditary C7 deficiency: diagnosis and HLA studies in a French-Canadian family, *J. Clin. Invest.*, 60, 1061, 1977.
15. **Lint, T.F., Osofsky, S.G., Nemerow, G.R., Tausk, K., and Gewurz, H.**, Inherited deficiency of the seventh component of complement (C7) in man: Description, relationship to HLA and partial characterization of an inhibitory activity, *Clin. Res.*, 24, 543A, 1976.
16. **Day, N.K., Degos, L., Beth, E., Sas Portes, M., Gharbi, R., and Giraldo, G.**, C8 deficiency in a family with xeroderma pigmentosum. Lack of linkage to the HL-A region, in *HLA and Disease*, INSERM, Paris, 1976, 197.
17. **Merritt, A.D., Petersen, B.H., Biegel, A.A., Meyers, D.A., Brooks, G.F., and Hodes, M.E.**, Chromosome 6: linkage of the eighth component of complement (C8) to the histocompatibility region (HLA), *Human Gene Mapping 3*, Bergsma, D., Ed., S. Karger, New York, 1976, 24.
18. **Teisburg, P., Olaisen, B., Gedde-Dahl, T., and Thorsby, E.**, On the localization of the Gb locus within the MHS region of chromosome no. 6., *Tissue Antigens*, 5, 257, 1975.
19. **Bajtai, G., Hernadi, E., and Ambrus, M.**, HLA-A1 and B8 antigens in selective IgA deficiency, in *HLA and Disease*, INSERM, Paris, 1976, 193.
20. **Berkel, A.I. and Ersoy, F.**, HLA antigens in ataxia-telangiectasia, in *HLA and Disease*, INSERM, Paris, 1976, 194.
21. **Hors, J., Griscelli, C., Schmid, M., and Dausset, J.**, HL-A antigens and immune deficiency states, *Br. Med. J.*, 4, 45, 1974.
22. **Buckley, R.H., MacQueen, J.M., and Ward, F.E.**, HLA antigens in primary immunodeficiency diseases, in *HLA and Disease*, INSERM, Paris, 1976, 195.
23. **Rimm, A. A., and Bortin, M. M.**, HLA antigens and severe combined immunodeficiency disease, *Lancet*, 1, 1361, 1977.
24. **Hagele, R., Evers, K.G., Leven, B., and Kruger, J.**, HLA frequencies in primary immunodeficiency diseases (IDD), in *HLA and Disease*, INSERM, Paris, 1976, 200.
25. **Van Camp, B., Dergent-Cole, Petermans, M.**, HLA-B7 and idiopathic paraproteinaemia, in *HLA and Disease*, INSERM, Paris, 1976, 213.
26. **Stiehm, E.R., Lawlor, G.J., Kaplan, M.S., Greenwald, H.L., Neerhout, R.C., Sengar, D.P.S., and Terasaki, P.I.**, Immunologic reconstitution in severe combined immunodeficiency without bone-marrow chromosomal chimerism, *N. Engl. J. Med.*, 286, 797, 1972.
27. **Terasaki, P.I.**, personal communication, 1976.

IV. CONNECTIVE TISSUE DISEASES

A. Systemic Lupus Erythematosus (SLE)

This disease (Table 4) was one of the first thought to have an HLA antigen association. In 1971, Grumet reported a study of 40 patients of different races in whom B8 was present in 33% and B15 in 40% compared to 16 and 10%, respectively, of controls.[1] These authors also reported four triplets, that is, three antigens in one series rather than the usual two, at the B locus which may have been a reflection of the difficulties with certain antisera at that point in time or possibly evidence of some additional HLA or even non-HLA antigens. Even though these results have never received strong confirmation in numerous studies since that time,[2-6] nevertheless, this paper made the important contribution of pointing out the necessity of taking into

TABLE 4

Systemic Lupus Erythematosus in Caucasians

HLA Ag	Controls with Ag (%)	Patients with Ag (%)	Patients with Ag (n)	Patients total (n)	Relative risk	Ref.
B8	16.0	36	9	25	3.0	1
	21.0	19.0	8	42	1.0	4
	23.1	30.8	20	65	1.5	3
	12.0	35.0	17	49	3.8	5
	17.4	37.8	17	45	2.9	6
	18.8	38.0	38	100	2.6	12
	—	48.0	24	50	3.6	13

account, for statistical purposes, the number of antigen specificities being tested. When one tests a large number of antigens in a blind fashion, one might erroneously conclude that significant associations were present when the association was due to nothing more than the fact that by looking at a large number of antigens, e.g., 20, by pure chance, one might show a significant deviation at the $P<0.05$ level. Part of the problem that supported the relative increase in B8 may have been the fact that the control series of 82 normal Caucasians had a frequency of B8 of only 16%, which was lower than the 21.2% these same authors reported in their controls used for the study of thyrotoxicosis.[7]

In 1971, another early study of 24 patients by Waters suggested a possible increase of a B15-related antigen, especially in the 15 patients whose disease onset was before 25 years of age.[8] However, in 1972, Arnett found essentially a normal distribution of HLA antigens in 40 patients though there was a slight increase in B13.[9] In the study of lupus by Nies which involved 124 individuals (42 Caucasians, 40 Black Americans, 40 Mexican Americans, and 2 Orientals) there was no significant difference found in any of the antigens with the exception of B5 in Black Americans ($P<0.01$).[4] Furthermore, there was no significant association between an HLA antigen and any clinical or laboratory expressions of the disease. Lymphocytotoxic antibody occurred in 36% of the patients and appeared to correlate with low total hemolytic complement and leukopenia.

In 1975, Kissmeyer-Nielsen reported that 65 Danish patients revealed no difference in HLA antigen frequency even when examined according to sex, age, or the ABO blood group. [3] Although B8 was increased to 31% in patients compared to 23% of controls, the difference was nonsignificant. B13 and B15 occurred with normal frequency. Furthermore, there was no confirmation of the occurrence of triplets. Leukoagglutinating antibodies were found in 1 of 11 patients tested, lymphocytotoxins in 4 of 59, and platelet complement fixing antibody in 13 of 55. The occurrence of triplets found by Grumet was explained by these authors as false positive reactions caused by the augmentation of weak typing sera by one of these antibodies or LE factors coating lymphocytes.

A recent large study of 120 patients (71 Blacks and 49 Caucasians) and 120 matched controls showed nonsignificant increases of A1 and B8, the former more prevalent in Blacks and the latter in Caucasians. [5] A1 was more frequent in those with an early disease onset, and A1, B8, and the A1-B8 phenotype were more frequent with SLE having renal and CNS involvement. However, the results in American Blacks were at variance with the increases in B7, B5, and Bw35 found by Bitter, [10] in A19 and B5 by Stastny, [11] and B5 by Nies. [4]

Even the most recent studies indicate only a modest increase in B8. Dostal studied

45 patients with SLE, 38 females and 7 males, and found a B8 frequency of 37.8% compared to 17.4% of 1200 healthy controls (corrected $P<0.05$ with a relative risk of 2.88) [6] His analysis of several studies from the literature showed a combined relative risk of 1.91 in a total of 467 SLE patients, thus supporting a weak but possibly important association between SLE and B8. [1,3-5,12] Berg-Loonen additionally reported that in 50 patients with systemic lupus, 24, or 48%, were positive for B8, with a relative risk of 3.64 for B8. [13] In 100 patients, Szegedi and Stenszky found an increased frequency of B5 (24% compared to 14.1% in 450 healthy controls), B8 (38% compared to 18.8%), and of the A1-B8 haplotype in patients from 28 families. [12]

Age and sex were found to be important in other studies. Sixty-six patients with discoid lupus erythematosus (DLE) showed an increased frequency of B7 in both young males and females with the onset of the disease between 15 and 39 years, and in B8 only in females over the age of 40 years. [14] In systemic lupus, B8 was more frequent at all ages. [14] In females between the ages of 15 and 39 years who presented with DLE that later transformed to SLE, a significant increase in B8 occurred which suggested that B8 had a particular contribution to the risk of lupus in females. [14]

In a small series of 17 patients with lupus nephropathy, Stefanova found that 66% of the patients had B15 and Aw24, which were 12 times and 4 times, respectively, more frequent than in controls (See Renal Disease [Chapter 7, Section VIII]).[15]

Three D-series specificities (Dw2, Dw3, and Dw4) were studied by Hanson in 47 Caucasian patients with SLE. [16] They noted a slight but insignificant increase in B8 (27.1% compared to 20.8% of their 48 matched controls) and B15 (10.4% compared to 6.2% of their matched controls). With respect to the D series specificities, Dw2 occurred in 10 of 47 (21.3% compared to 12.5% of controls), Dw3 in 8 of 47 (17.0% compared to 14.6% of controls), and Dw4 in only 2 of 47 patients (4.2% compared to 8.3% of controls), none of which was significant.

In a study of the possible association between C2 deficiency and HLA genes in both systemic lupus and discoid lupus, Agnello found, in a kindred of 40 patients of whom 14 of 38 (37%) of the homozygous and 11 of 136 (6.6%) of the heterozygous individuals had SLE or DLE, that 18 of the 19 individuals typed had the C2 deficiency usually associated with the A10-B18 haplotype, and 7 of 8 patients tested had Dw2. [17] Patients with C2 deficiency and SLE appeared to have a different immunologic pattern in that 5 of 8 homozygous C2-deficient lupus patients had only weak or absent antibodies to native DNA, only 2 of 9 had renal disease, and 5 of 7 had no immunoglobulins or complement deposited in the skin as did 3 of 4 with discoid lupus. Thus, the absence of SLE in more than 60% of homozygous C2-deficient individuals and the lack of an increased frequency of the A10-B18 haplotype among SLE patients indicated that neither gene alone predisposed to SLE. Furthermore, the combination of C2 deficiency and its typical HLA gene products (A10-B18) or a yet unidentified gene led to a form of SLE which lacked the immune responses usually seen in classical SLE.

B. Progressive Systemic Sclerosis (PSS) (Scleroderma)

In a study by Birnbaum of 106 patients with PSS, 53 of whom had the classic diffuse form and 53 the so-called "CREST" syndrome (calcinosis, Raynaud's phenomenon, esophageal dysfunction, sclerodactylia, and telangiectasia), which is generally a milder form with only limited skin involvement, no significant deviation in antigen frequencies could be found in the group as a whole, in the diffuse, or in the "CREST" syndrome group.[18] This finding in a large group of patients confirmed the same group's earlier results.

Similarly, a French study of 49 patients also found no association with any HLA antigen.[19] However, a recent study of 40 patients with PSS detected an increase of both A9 (38%) and Aw24 (30%), compared to 14% and 11%, respectively, of controls.[20]

TABLE 5

Sjögren's Syndrome

HLA Ag	Controls with Ag (%)	Patients with Ag (%)	Patients with Ag (n)	Patients total (n)	Relative risk	Ref.
B8	31.2	38.1	8	21	1.4	21
	21.0	58.3	14	24	5.2	22
	22.0	52.8	19	36	3.9	23
	22.1	45.0	21	47	2.9	25
	22.1	62.5[a]	10	16	5.9	25
	21.0	43.0	27	63	2.8	27
	18.9	58.6	17	29	4.0	28
Dw3	17.0	58.0	14	24	6.9	27
	10.3	69.0	20	29	19.2	28

[a] With PM or PSS.

C. Sjögren's Syndrome

The first antigen deviation found in Sjögren's syndrome (Table 5) is B8 (combined uncorrected P = 1.9×10^{-7}) with a relative risk (RR) of 3.25 and a 95% confidence interval of 2.08 to 5.05.[2] This syndrome is based on the combination of keratoconjunctivitis sicca, xerostomia, and rheumatoid arthritis and is applied when any two of these features are present. However, in some cases, lupus erythematosus, scleroderma, or polyarteritis may replace rheumatoid arthritis in the complex. Nearly all patients show positive tests for rheumatoid factor, but a high percentage of patients also have positive tests for antinuclear antibody, thyroglobulin antibody, and complement fixing antibodies to various organs and tissues. Some patients may go on to develop reticulum cell sarcomas.

The initial report of HLA antigen frequencies and Sjögren's syndrome was particularly aimed at the sicca syndrome in which only keratoconjunctivitis sicca and xerostomia are present without apparent joint disease.[21] Those 21 patients showed no significant alteration of any HLA antigen frequencies, though B8 was slightly increased in frequency from 31.2% in the control population to 38.1% in the patients. However, subsequent studies of the complete syndrome by Gershwin[22] and Ivanyi[23] both showed that in a total of 60 patients, the antigen frequency of B8 was significantly increased. Gershwin's explanation for the finding of a positive association in contrast to Sturrock's earlier negative study was that his own patients' younger age enhanced the correlation with a particular antigen.[22] Furthermore, Gershwin's control frequency of B8 was lower than Sturrock's (21% compared to 31%) and 23 of the 24 patients were female, whereas the sex distribution in the other study was not noted.

This disease fits into the group characterized by autoantibodies, a tendency to occur in young females, and an association with B8, though rheumatoid arthritis is a notable exception to this. In a study of 36 patients with Sjögren's syndrome, Ivanyi also found a significant increase in the B8 antigen but no association between the antigen and rheumatoid factor, which was present in 11 of the 33 patients tested for this factor.[23] They also noted that all their patients with Sjögren's syndrome associated with rheumatoid arthritis were B8-negative. The syndrome in B8-positive patients might be different from that in B8-negative patients with rheumatoid arthritis, a concept supported by the finding that different precipitating antigen-antibody systems existed in patients with Sjögren's syndrome depending on the presence or absence of rheumatoid arthritis.[24] In contrast to the negative association with B8 in patients positive for rheumatoid factor, 5 of 9 patients with rheumatoid factor had B7. Further support for the type of

B8 association in Sjögren's syndrome came from more recent studies by Clough in 47 patients with Sjögren's syndrome, 2l of whom had B8 (45% compared to 22% of 485 Caucasian controls).[25] The most interesting aspect of this study was the fact that when rheumatoid arthritis was not present and polymyositis (PM) or progressive systemic sclerosis (PSS) was, the frequency of the B8 antigen increased. When PM was present, 5 of 7 patients (71%) had B8, and when PSS was present, 5 of 9 (56%) had B8, whereas when rheumatoid arthritis was present, only 11 of 31 (35%) had the antigen. Furthermore, when Sjögren's syndrome was associated with systemic lupus erythematosus or with necrotizing vasculitis, just as with rheumatoid arthritis, there was no statistically significant increase in B8. No correlation with sex, age of onset, or complications was detected. Thus, this study suggested that the presence of B8 in individuals with Sjögren's syndrome was associated with those rheumatic diseases such as PM and PSS in which cellular rather than humoral immune mechanisms play the largest role. This clinical finding is paralleled by in vitro studies which showed heightened reactivity in mixed lymphocyte cultures with cells from B8-positive donors.[26]

D-series antigens have been studied by two groups.[27-29] Opelz found a significant association with Dw3 in 24 patients with Sjögren's disease.[27] Fifty-eight percent of Sjögren's syndrome patients compared to 17% of 59 control patients had the Dw3 antigen. Of 63 patients with Sjögren's syndrome typed for B-locus antigens, 27 or 43% had B8, and in the subgroup of 24 patients typed for D-locus antigens, 2 of 12 patients who lacked B8 had Dw3. Further data on Dw3 indicate that its association, similar to Clough's finding for B8,[25] is primarily with Sjögren's syndrome that lacks rheumatoid arthritis.[29] Very recent studies of B lymphocyte antigens in 24 patients with the sicca syndrome have revealed two sera that reacted with all patients and only 24 to 37% of controls.[30] It was suggested that these sera detected two Ia antigens, the gene products of two immune response loci necessary for development of the disease.

4. Dermatromyositis

Twelve of sixteen patients with juvenile dermatomyositis had B8 (75% compared to 21% of controls) ($P < 0.0005$), yielding a relative risk of 11.5.[31]

REFERENCES

1. **Grumet, F.C., Coukell, A., Bodmer, J.G., Bodmer, W.F., and McDevitt, H. O.,** Histocompatibility (HL-A) antigens associated with systemic lupus erythematosus, *N. Engl. J. Med.*, 285, 193, 1971.
2. **Ryder, L.P. and Svejgaard, A.,** Associations between HLA and disease, Report from the HLA and Disease Registry of Copenhagen, 1976.
3. **Kissmeyer-Nielsen, F., Kjerbye, K.E., Andersen, E., and Halberg, P.,** HL-A antigens in systemic lupus erythematosus, *Transplant. Rev.*, 22, 164, 1975.
4. **Nies, K.M., Brown, J.C., Dubois, E.L., Quismorio, F.P., Friou, G.J., and Terasaki, P.I.,** Histocompatibility (HL-A) antigens and lymphocytotoxic antibodies in systemic lupus erythematosus (SLE), *Arthritis Rheum.*, 17, 397, 1974.
5. **Goldberg, M.A., Arnett, F.C., Bias, W.B., and Shulman, L.E.,** Histocompatibility antigens in systemic lupus erythematosus, *Arthritis Rheum.*, 19, 129, 1976.
6. **Dostal, C., Ivanyi, D., Macurova, H., Hana, I., and Strejcek, J.,** HLA-B8 antigen in systemic lupus erythematosus (SLE), in *HLA and Disease,* INSERM, Paris, 1976, 199.
7. **Grumet, F.C., Payne, R.O., Konishi, J., and Kriss, J.P.,** HL-A antigens as markers for disease susceptibility and autoimmunity in Graves' disease, *J. Clin. Endocrinol. Metab.*, 39, 1115, 1974.
8. **Waters, H., Konrad, P., and Walford, R.L.,** The distribution of HL-A histocompatibility factors and genes in patients with systemic lupus erythematosus, *Tissue Antigens,* 1, 68, 1971.
9. **Arnett, F.C., Bias, W.B., and Shulman, L.E.,** HL-A in systemic lupus erythematosus (SLE), *Arthritis Rheum.*, 15, 428, 1972.

10. **Bitter, T., Mottironi, W.D., and Terasaki, P.I.,** HL-A antigens associated with lupus erythematosus, *N. Engl. J. Med.,* 285, 435, 1972.
11. **Stastny, P.,** The distribution of HL-A antigens in black patients with systemic lupus erythematosus (SLE), *Arthritis Rheum.,* 15, 455, 1972.
12. **Szegedi, G. and Stenszky, V.,** HLA and systemic lupus erythematosus, in *HLA and Disease,* INSERM, Paris, 1976, 292.
13. **Berg-Loonen, E.M. Van Den, Swaak, A.J.G., Nijenhuis, L.E., Feltkamp, T.E.W., and Engelfreit, C.P.,** Histocompatibility antigens and other genetic markers in patients with systemic lupus erythematosus, in *HLA and Disease,* INSERM, Paris, 1976, 214.
14. **Millard, L.G., Rowell, N.R., and Rajah, S.M.,** Histocompatibility antigens in discoid and systemic lupus erythematosus, in *HLA and Disease,* INSERM, Paris, 1976, 112.
15. **Stefanova, G.,** Relationship between HLA and other immunological tests in nephropathy due to SLE, in *HLA and Disease,* INSERM, Paris, 1976, 211.
16. **Hansen, J.A., Rothfield, N.F., Bobrone, M., Jersild, C., Wernet, P., and Dupont, B.,** MLC determinants (HLA-D) in patients with systemic lupus erythematosus (SLE), in *HLA and Disease,* INSERM, Paris, 1976, 201.
17. **Agnello, V.,** Association of C2 deficiency (C2D) and HLA genes with systemic and discoid lupus erythematosus (SLE, DLE), in *HLA and Disease,* INSERM, Paris, 1976, 296.
18. **Birnbaum, N.S., Rodnan, G.P., Rabin, B.S., and Bassion, S.,** Histocompatibility antigens in progressive systemic sclerosis (scleroderma), *J. Rheumatol.,* 4, 425, 1977.
19. **Crouzet, J., Marbach, M.C., Camus, J.P., Godeau, P., Herreman, G., Richier, D., Hors, J., and Dausset, J.,** Recherche d'une association entre antigenes HL-A et sclerodermie systemique, *Nouv. Presse Med.,* 4, 2489, 1975.
20. **Clements, P. J., Opelz, G., Terasaki, P. I., Mickey, M. R., and Furst, D.,** Association of HLA antigen A9 with progressive systemic sclerosis (scleroderma), *Tissue Antigens,* 11, 357, 1978.
21. **Sturrock, R.D., Canesi, B.A., Dick, H.M., and Dick, W.C.,** HL-A antigens and the sicca syndrome, *Ann. Rheum. Dis.,* 33, 165, 1974.
22. **Gershwin, M.E., Terasaki, P.I., Graw, R., and Chused, T.M.,** Increased frequency of HL-A8 in Sjögren's syndrome, *Tissue Antigens,* 6, 342, 1975.
23. **Ivanyi, D., Drizhal, I., Erbenova, E., Horejs, J., Salavec, M., Macurova, H., Dostal, C., Balik, J., and Juran, J.,** HL-A in Sjögren's syndrome, *Tissue Antigens,* 7, 45, 1976.
24. **Alspaugh, M.A. and Tan, E.M.,** Antibodies to cellular antigens in Sjögren's syndrome, *J. Clin. Invest.,* 55, 1067, 1975.
25. **Clough, J.D., Aponte, C.J., and Braun, W.E.,** HLA-B8 and clinical features of Sjögren's syndrome, *Clin. Immunol. Immunopathol.,* in press.
26. **Osoba, D. and Falk, J.,** HLA genes regulating the magnitude of the mixed lymphocyte reaction (MLR), *Fed. Proc.,* 35, 712, 1976.
27. **Opelz, G., Terasaki, P., Vogten, A., Schalm, S., Summerskill, W., Kassan, S., Chused, T., and Fye, K.,** Association of HLA-DW3 with chronic active liver disease and Sjögren's disease, in *HLA and Disease,* INSERM, Paris, 1976, 167.
28. **Hinzova, E., Ivanyi, D., Sula, K., Horejs, J., Dostal, C., Drizhal, I.,** HLA-Dw3 in Sjögren's syndrome, *Tissue Antigens,* 9, 8, 1977.
29. **Chused, T.M., Kassan, S.S., Opelz, G., Montsopoulos, H. M., and Terasaki, P.I.,** Sjögren's syndrome associated with HLA- Dw3, *N. Engl. J. Med.,* 296, 895, 1977.
30. **Montsopoulos, H. M., Chused, T. M., Johnson, A. H., Knudsen, B., and Mann, D. L.,** B lymphocyte antigens in sicca syndrome, *Science,* 199, 1441, 1978.
31. **Pachman, L. M., Jonasson, O., Cannon, R. A., and Friedman, J. M.,** HLA-B8 in juvenile dermatomyositis, *Lancet,* 2, 567, 1977.

V. DERMATOLOGY

A. Psoriasis Vulgaris

The initial antigens shown to occur with increased frequency in this disease are B13, B17, B37, as well as Bw16 in unspecified types of psoriasis. These antigens have combined uncorrected P values of $< 1.0 \times 10^{-10}$, $< 1.0 \times 10^{-10}$, 7.3×10^{-8}, and 2.1×10^{-6}, and relative risks of 4.65, 4.90, 6.35, and 4.24, respectively.[1] The major studies in this disease are shown in Table 6.

TABLE 6

Psoriasis in Caucasians and Japanese

HLA Ag	Controls with Ag (%)	Patients with Ag (%)	Patients with Ag (n)	Patients total (n)	Relative risk	Ref.
			Caucasians			
B13						
Unspecified	3.4	27.3	12	44	9.5	2
	4.9	14.7	23	156	3.3	3
	1.0	6.9	6	87	6.3	16
	7.9	23.1	24	104	3.5	17
	4.3	12.7	14	110	3.2	6
	5.2	7.2	9	125	1.4	18
	7.1	8.9	9	101	1.3	5
Vulgaris	2.8	21.1	16	76	9.0	19
	4.2	15.0	12	80	4.1	19
	8.0	24.0	30	125	3.6	14
B17						
Unspecified	9.0	22.7	10	44	2.9	2
	8.0	26.3	41	156	4.1	3
	0.5	10.3	9	87	15.9	16
	7.9	21.2	22	104	3.1	17
	3.9	21.8	24	110	6.6	6
	6.0	50.4	63	125	15.4	18
	6.2	30.7	31	101	6.3	5
Vulgaris	8.2	27.6	21	76	4.3	19
	8.0	36.3	29	80	6.6	19
	4.0	18.4	23	125	5.3	14
Bw16						
Unspecified	2.9	10.6	11	104	3.9	17
	5.3	21.8	22	101	4.7	5
B37						
Vulgaris	1.0	6.4	8	125	6.6	14
			Japanese			
A1						
Vulgaris	1.5	16.6	9	54	13.0	13
B37						
Vulgaris	4.5	20.4	11	54	5.4	13

The earliest HLA studies of psoriasis were published simultaneously by Russell and White.[2,3] Russell found that in 44 unrelated patients with psoriasis, B13 was present in 27.3% and B17 in 22.7% compared to 3.4 and 9.0% of controls, respectively.[2] Of the 12 patients with B13, 10 developed psoriasis before the age of 33 years. In 22 members of a family with psoriasis also studied by this group, A1 and B17 were present in family members who had a personal history of psoriasis as well as in most members who did not. The absence of psoriasis in subjects having A1-B17 is not unexpected in view of twin studies in psoriasis that showed incomplete concordance, indicating the contribution of environmental factors.

In a companion report, White found that in 156 psoriatic patients, B13 was present in 14.7% and B17 in 25.7% compared to 5 and 8% of controls, respectively.[3] When a subgroup of 42 healthy subjects with Jewish surnames were found to have A1, A10, and B17 in a higher frequency than in healthy controls with non-Jewish surnames, the contribution of race to the alteration in antigen frequency in this group was investigated further. It was then found that the increased frequencies of A1 and A10 in psoriasis disappeared when the Jewish patients were removed from the study. Furthermore, a remarkable parallelism was shown by normal Jewish controls in the specificities that were found to be skewed in psoriasis as shown by the fact that no statistically significant difference could be shown between healthy Jewish subjects and Jewish patients with psoriasis. This point emphasizes the necessity for carefully evaluating the racial background of any disease being studied and using as the appropriate control the same race without the disease as is being investigated with the disease. A large family study by White showed that all affected members had the A1-B17 haplotype but two members as yet unaffected by the disease also had the A1-B17 haplotype. The other interesting feature of this study was the fact that B12 was found to be significantly decreased in this population, a finding which is reminiscent of the decreased frequency of B12 in patients with multiple sclerosis. [4] These authors could find no correlation of any HLA specificity with the clinical features of the disease such as arthritis, severity, distribution of the lesions, sex, and red cell type, although those with B17 appeared to have a higher rate of affected relatives and a slightly earlier age of onset, findings later confirmed by Krulig. [5]

A later study by Seignalet, while confirming the fundamental observations of the initial studies that B13 and B17 were significantly increased in psoriatics and that there was no relationship between the HLA antigens and articular disease, noted that in four families of psoriatics, the disease was transmitted independently of the B17 antigen in two families, in a third family, followed an A1-B17 haplotype in one group of individuals and another set of haplotypes without B17 in other members of the same family, and in the fourth family, was transmitted with B17.[6] Thus, psoriasis was not always transmitted with B17 as suggested by Russell's and White's studies.

Seignalet also found an increase in IgE titer and an anti-IgG activity demonstrable on peripheral blood lymphocytes.[6] The fact that hypersensitivity reactions of the skin are associated with histamine release, mediated by IgE, and inhibited by cyclic AMP is placed in a more relevant perspective by the finding that cyclic AMP levels may be determined by Ia-like antigens in the major histocompatibility complex in animal systems.[7]

In more recent studies, another antigen, Bw16, has been found to be significantly increased. Krulig found that Bw16 occurred in 22% of 101 psoriatic patients compared to 5% of controls as well as finding the well-known increase in frequency of B17, 31% compared to 6% of controls. [5] On the other hand, B13 was only slightly increased to 9% of 101 patients compared to 7% of controls. Krulig's study made some further interesting observations on the clinical associations with certain of these antigens. For example, patients with B17 had an earlier age of onset (22 years), whereas those with either B17 or Bw16 or both had a more severe form of the disease involving greater than 50% of the body surface as well as the nails.

However, several different clinical associations were noted in the dermatology workshop summary of the 1976 Paris Symposium on HLA and Disease.

1. Patients with B37 were the ones likely to develop psoriasis at a very early age, even in infancy.
2. B17 was much more frequent in those families where there was a very high incidence of psoriasis.
3. No confirmation of Krulig's data could be found relating to the severity of the disease.[8]

It has been observed that the American Indian does not suffer from psoriasis and lacks the gene for B13. [8] Similarly, the Black African, who has only about a 1% frequency of B13 but about a 21% frequency of B17, rarely has psoriasis.

A C-series antigen, Cw6, may have the most significant association with psoriasis, but it has not been widely tested.[9]

In a study of 7 HLA-D alleles in psoriatic patients, Grosse-Wilde found in 68 patients that Dw2 had a frequency of 30.8% compared to 14% in normal controls.[10] Another D antigen, called EI, was approximately three times greater in psoriatic patients than in controls (23.1% compared to 8.1%). No other D-antigen deviations were found. However, it was noted by these authors that EI had a strong linkage disequilibrium with both B13 and B17, possibly accounting for its higher frequency in the psoriatic patients.

Möller's studies of psoriatic families indicated that there were at least two genes responsible for the increased risk of the disease, both for the skin manifestations and the arthritis; that they were both in the major histocompatibility complex within the D-locus vicinity; and that they may occur in the *cis* position on the same haplotype or in the *trans* position on different haplotypes.[11] This concept derived support from the families first studied by Russell and White and was foreshadowed by Svejgaard's family analyses.[12]

In 54 unrelated Japanese patients with psoriasis vulgaris, the distribution of antigen deviations was different from that in Caucasians in that A1 and B37 were increased to 16.6 and 20.4% compared to 1.5 and 4.5% of controls, respectively, and B13 and B17 were not increased significantly.[13] Karvonen also found an increase of B37 in Caucasian psoriatic patients, particularly those whose disease began before 40 years of age.[14]

In examining the correlation of HLA antigens with psoriasis, it is important to look at the pathologic events in this disease. The predominant event in chronic psoriasis vulgaris is an increased epidermopoiesis with an epidermal turnover rate nine times that of normal.[8] The second pathologic event is the migration of neutrophils from the subepidermal venules across the epidermis into the stratum corneum. This cell migration occurs in all forms of psoriasis, but in some acute flares of psoriasis, the epidermopoiesis is minimal and the overwhelming event is the chemotaxis of neutrophils with pustular psoriasis as the end result. It should be emphasized that pustular psoriasis has no HLA antigen association.[12,13,15] Consequently, the altered HLA antigen frequencies may be associated with pathologic events leading to the increased epidermopoiesis, a reflection of cyclic AMP activity, which in turn may be controlled by an Ia antigen as suggested by animal studies.[7] This and Seignalet's IgE data both involve the activity level of cyclic AMP as a cause of psoriasis, which may be regulated by a gene in the major histocompatibility complex.[7]

B. Dermatitis Herpetiformis

The first antigen shown to occur with increased frequency in this disease is B8 (combined uncorrected $P < 1.0 \times 10^{-10}$) with a relative risk (RR) of 9.23 and 95% confidence interval of 6.74 to 12.63.[1] The major studies in this disease are shown in Table 7A.

Dermatitis herpetiformis (DH) is a bullous dermatologic disease frequently associated with duodenal and jejunal villous atrophy similar to that found in gluten sensitive enteropathy. It is associated with circulating autoantibodies, quantitative alterations in serum immunoglobulins, circulating immune complexes, and IgA deposits in the skin. Because of the similarities to gluten sensitive enteropathy, especially the fact that some patients with DH have a reversion of the small-intestinal abnormalities to normal after withdrawal of gluten from the diet, a possible association with HLA was sought like that of GSE with B8. Studies by Katz,[20] White,[21] Gebhard,[22] Solheim,[23] Reunala,[24]

TABLE 7A

Dermatitis herpetiformis

HLA Ag	Controls with Ag (%)	Patients with Ag (%)	Patients with Ag (n)	Patients total (n)	Relative risk	Ref.
B8	23.9	57.7	15	26	4.3	20
	33.1	60.0	21	35	3.0	21
	17.5	86.9	53	61	29.5	24
	26.1	78.9	30	38	10.1	25
	22.1	64.0	16	25	6.3	27
	24.9	79.5	35	44	11.3	23
	23.7	85.0	34	40	18.2	26
Dw3	19.0	75.0	15	20	12.7	23
	19.7	92.5	37	40	50.1	26

Seah,[25] Thomsen,[26] and Roberts-Thomson[27] have all shown an increased frequency in the B8 antigen. Once again, the increased frequency in A1 was attributable to the linkage disequilibrium that it has with B8.

In the initial study by Katz, 58% of 26 patients with dermatitis herpetiformis were found to have B8 compared to 24% of normal controls, a result that resembled their finding of an 88% B8-antigen frequency in patients with gluten-sensitive enteropathy.[20] When treatment was undertaken with gluten-free diets and the intestinal lesions in DH were corrected, the skin eruption usually did not respond, and conversely, effective treatment of the skin disease did not reverse the intestinal abnormality. The skin lesions in DH show subepidermal blisters with an acute inflammatory infiltrate of neutrophils in the dermal papillae adjacent to the blister. The fact that these skin lesions appear to have no relationship to the B8-antigen association is reminiscent of the finding in psoriasis in which the pustular form, also with heavy neutrophilic infiltration, has no HLA antigen association, whereas the vulgaris form with increased epidermopoiesis does. The authors speculated on possible causes common to DH and GSE for these associations occurring with B8. One possibility was that of a fortuitously linked immune response gene that might determine the response to a challenge with gluten in DH and GSE, manifested by IgA deposition in the skin in the former and IgA and IgM mucosal synthesis in the latter disease. Another possibility considered was that the HLA antigens on the cell surface could act as receptors capable of binding an infectious agent or other materials of pathologic significance such as gluten or its alpha gliaden fragments, as described under GSE (Chapter 7, Section VII.I).[28] Here it was pointed out that B8 alone could not be the sole factor in the pathogenesis of these diseases because of its high frequency in the normal population and the significant proportion of patients with the disease who lack the antigen.[28]

Despite the clinical and histologic similarities of the intestinal lesions in GSE and DH and the superficially satisfying answer that B8 would offer as a common factor in these diseases, there are conflicting data concerning a higher frequency of B8 in DH with compared to DH without intestinal lesions. In 28 patients with DH studied by Gebhard, 84% of patients with associated villous atrophy as determined by small bowel biopsy had B8, whereas only 33% of patients without gastrointestinal disease had B8.[22] However, a contemporary study by White revealed that in 23 patients with abnormal jejunal biopsies and dermatitis herpetiformis, the frequency of B8 was only slightly higher in that group (65.5%) compared to 50 % in the eight cases with normal jejunal biopsies, a nonsignificant difference.[21] These authors pointed out the fact that jejunal biopsy material must be carefully investigated because a multiple biopsy technique revealed that jejunal changes were present in almost 100% of patients with DH and not in about 66% as had originally been suspected. Another study of 27 patients with dermatitis herpetiformis and a 64% frequency of B8 also revealed no correlation of the small bowel lesions with B8 but rather an inverse correlation with age.[27] Eighteen of these subjects with morphologic small bowel changes had a mean age of 38 years, which was significantly lower ($P < 0.001$) than the nine subjects with normal small bowel whose mean age was 60 years.[27] An age-related immune responsiveness was offered as a cause for the prevalence of the GI lesions in the younger patients. The production of autoantibodies by DH patients with and without B8 has not been found to be significantly different, since they occurred in 10 of 21 B8-positive patients and 7 of 14 B8-negative.[21]

D-series antigens were investigated by Thomsen in 40 patients with dermatitis herpetiformis. Of the 23 patients with celiac disease in the presence of dermatitis herpetiformis, 87% had B8, whereas 95.7% had LD 8a, now known as Dw3, and 34.8% had LD 12a. [26] The relative risk with each of these antigens was 21.7, 89.4, and 5.9, respectively. In the 15 patients who did not have celiac disease with dermatitis herpetiformis, the frequency of B8 was 80%; of 8a or Dw3, 86.7%; and of 12a, 13.3%, with relative risks of 12.8, 26.4, and 1.7, respectively.

In addition to the fact that in dermatitis herpetiformis, the frequency of Dw3 was increased to a greater extent than was B8, the authors also made the point that they could find no difference in the frequency of either B8 or Dw3 in those patients with or without enteropathy, again contradicting the findings of Gebhard[22] and supporting the findings of White[21] and Roberts-Thomson[27] concerning B8. The finding of an increase in the LD 12a antigen was somewhat surprising. It suggested a dual mechanism in the pathogenesis of the enteropathy associated with dermatitis herpetiformis in view of the fact that others had found an increase in GSE of the B-cell antigen in weak linkage disequilibrium with this D-locus antigen, namely B12.[29] Similar results were obtained by Solheim, who found that in 34 patients, 79% had B8 compared to 24.9% of controls and that in the 20 patients who were typed for D-series antigens, 75% had Dw3 compared to 19% of 136 controls.[23]

Using serologic techniques for the definition of possible DR antigens, Keuning, Mann, and Thorsby each found an increased frequency of a B cell antigen in patients with dermatitis herpetiformis.[30-32]

C. Pemphigus Vulgaris

The primary antigen shown to occur with increased frequency in this disease is A10 (combined uncorrected $P < 1.2 \times 10^{-4}$) with a relative risk (RR) of 3.14.[1] The major studies in this disease are shown in Table 7B.

In pemphigus vulgaris characterized by acantholysis and a positive Nikolsky's sign, a first series antigen, A10, has received the most attention with a moderately increased frequency of 20% in non-Jewish individuals and strikingly increased frequency of 61% in Jewish patients. [33,34] Such a dramatic ethnic difference indicates the need for appropriate ethnic controls as was done so well by White in psoriasis.[3] Katz also reported an increase in B13 to 28% of patients compared to 4% of controls, but this study involved all forms of pemphigus in only 18 patients.[33]

D. Bullous Pemphigoid

Twenty unrelated Caucasian patients with bullous pemphigoid showed essentially no significant alterations in the frequency of HLA antigens.[35] The normal frequency of B8 in this disease, which has clinical features similar to dermatitis herpetiformis, is an important negative finding, however, and may reflect different pathogenic mechanisms.

TABLE 7B

Pemphigus vulgaris

HLA Ag	Controls with Ag (%)	Patients with Ag (%)	Patients with Ag (n)	Patients total (n)	Relative risk	Ref.
A10	15.5	22.2	4	18	1.7	33
Jewish	20.2	60.7	17	28	5.9	34
Non-Jewish	11.0	20.0	3	15	2.2	34

E. Vitiligo

Ninety unrelated Caucasian patients studied by Retornaz showed no significant deviation in any HLA antigen. [36] In 13 of the 57 patients tested, antithyroid antibodies were found and 4 of these patients had B13. This finding suggested that vitiligo was more a syndrome than a disease and that B13, though uninformative for the skin lesion, might help to select for those who had additional immunologic abnormalities.

F. Lichen Planus

Forty-three adult patients showed no significant increase in any HLA antigen although A3 was present in 37% (compared to 23% of controls), B8 in 26% (compared to 11% of controls), and B5 in 21% (compared to 13% of controls). [37] Even when the diseases were examined according to age of onset, sex of the patients, and extent of the lesion, no significant antigen association was found. Furthermore, no circulating antibodies were found by indirect immunofluorescence using epidermal nuclear muscle and rabbit vascular tissues.

G. Ehlers-Danlos Syndrome

A study of a single family of seven members, three of whom had Ehlers-Danlos syndrome, suggested an association between the disease and an HLA haplotype having an A series blank and B13.[38]

H. Keloids and Hypertrophic Scars

Twenty-five patients with keloids and/or hypertrophic scars were found to have an increased frequency of B14 (25% compared to 4% of 131 controls, with a relative risk of 6.3) and Bw16 (31% compared to 6% controls, with a relative risk of 4.8).[39]

I. Alopecia Areata

Two studies of this disease arrived at different conclusions. Kianto found a significant increase in B12 among 47 unrelated patients with this disease.[40] B12 occurred in 38% compared with 15% of 326 controls ($P < 0.01$). However, Kuntz found no significant elevation in any HLA antigen in 70 patients with the same disease.[41]

J. Hailey-Hailey's Disease

Hailey-Hailey's disease, an inheritable acantholytic disease, was studied in nine unrelated patients, four of whom had B8 and three of whom each had B5 and B40.[42] This series is too small to draw any conclusions, but it does indicate that the suspected "psoriatic" HLA antigens, namely B13, B17, Bw16, and B37, did not occur with unusual frequency.

K. Mycosis Fungoides

This cutaneous disease, which is believed to represent a slowly evolving T-cell neoplasm, was studied in 15 patients in the early stages of the disease.[43] A high frequency of both B8 (46.8%) and of the antigens in the Aw19 complex was observed, specifically Aw31 and Aw32, which together comprised 40% of the A-series antigens in the group.

L. Xeroderma Pigmentosum

This autosomal recessive disease, characterized by abnormal repair of sunlight-derived UV-radiation damage to the skin and eventually multiple malignancies, was studied in 16 families with 37 patients and 108 relatives. No significant deviations in HLA antigen frequencies were found.[44]

M. Recurrent Oral Ulcerations

This disease of unknown etiology was studied in 100 patients in whom the antigen B12 was found in 43% compared to 22% of controls, yielding a relative risk of 2.3.[45] In addition, the haplotype, A2-B12, occurred in 30% of these patients compared to 11% of controls, yielding a relative risk of 3.5. The detection of B12 might help in the differential diagnosis of this disease from Behçet's disease, which is associated with B5.

N. Behçet's Disease

The primary antigen shown to occur with increased frequency in this disease (Table 8) is B5 (combined uncorrected $P < 8.1 \times 10^{-5}$) with a relative risk (RR) of 4.29 and 95% confidence interval of 2.08 to 8.85.[1]

The increased frequency of B5 has been apparent not only in Caucasians but also in Japanese.[46-52] Unfortunately, some of the studies are quite small and the clinical disease is not always well defined. However, in 26 patients with the complete form of Behçet's disease (chronic relapsing oral and genital ulcerations, skin lesions, arthritis, uveitis with hypopion, and thrombophlebitis) the frequency of B5 was found to be 84.6% compared to 25.3% in 138 controls.[49]

The largest study thus far of 73 patients (27 males and 46 females) confirmed the importance of B5 that occurred in 88% of 17 patients with the complete syndrome, 59% of 37 with the incomplete, 25% of 8 with probable, and 36% of 11 with possible Behçet's.[53] B5 was greater in males in all forms of the disease. Family studies showed that, although the disease was typically associated with an HLA haplotype, the haplotype did not always contain B5.[53]

Another large study of 39 Caucasian patients with careful clinical delineation of the disease showed that B5 occurred in 54% of the patients compared to 13% of 591 controls, yielding a relative risk of 7.9.[50] Both of these studies confirmed Ohno's work that first showed an increase in the B5-antigen frequency and suggested that the presence of the antigen may be helpful in establishing the diagnosis in early cases.[46,47] Ohno also suggested that the suspected infectious etiology of Behçet's syndrome from Chlamydia would be extremely difficult and complex to prove because, although his patients with Behçet's did have higher antibody titer to Chlamydia, there was no difference in the frequency of B5 between the patients with and without the Chlamydial antibody.

O. Atopic Dermatitis

A3 and A9 each occurred in 15 of 45 patients (33%) with classical atopic dermatitis compared to 19 and 21%, respectively, in 870 controls (corrected $P < 0.05$ for each).[56] A study of 27 unrelated children with atopic dermatitis by Goudemand (16 cases of isolated disease and 11 associated with asthma) showed a significant increase in Bw35 (35% compared to 11.8% of 325 controls, corrected $P < 0.01$).[57]

In a study of 103 patients from Germany, no significant deviations were found in

TABLE 8

Behçet's Disease in Caucasians and Japanese

HLA Ag	Controls with Ag (%)	Patients with Ag (%)	Patients with Ag (n)	Patients total (n)	Relative risk	Ref.
			Caucasians			
B5	10.1	18.2	4	22	2.1	54
	16.3	37.3	3	8	3.2	48
	13.0	54.0	21	39	7.9	50
	25.3	84.6	22	26	16.2	49
			Japanese			
B5	30.8	75.0	33	44	6.5	47
	—	66.7	30	45	3.2	55

the HLA antigen frequencies or in the HLA haplotype frequencies of the patient population.[58] These authors investigated the possibility that the disease might segregate with an HLA haplotype in a family without exhibiting any significant association with a particular HLA antigen or a particular haplotype. They were able to show that there was a very unlikely chance ($P < 0.000058$) that the HLA haplotypes observed with atopic dermatitis in a set of 19 families would have occurred by chance alone. Therefore, they concluded that the susceptibility to atopic dermatitis was in some way related to the HLA system, though not discernible by any currently defined antigens.

P. Recrudescent Herpes Labialis (see Infectious Diseases [Chapter 7, Section VIII])

Q. Leprosy (see Infectious Diseases [Chapter 7, Section VIII])

R. Porphyria Cutanea Tarda

Five of six affected family members had A3 as did four of nine unrelated patients with porphyria cutanea tarda.[59] Thus, two diseases with abnormal iron metabolism, porphyria cutanea tarda and idiopathic hemochromatosis, are both associated with A3.

REFERENCES

1. **Ryder, L.P. and Svejgaard, A.**, Associations between HLA and disease, Report from the HLA and Disease Registry of Copenhagen, 1976.
2. **Russell, T.J., Schultes, L.M., and Kuban, D.J.**, Histocompatibility (HL-A) antigens associated with psoriasis, *N. Engl. J. Med.*, 287, 738, 1972.
3. **White, S.H., Newcomer, V.D., Mickey, M.R., and Terasaki, P.I.**, Disturbance of HL-A antigen frequency in psoriasis, *N. Engl. J. Med.*, 287, 740, 1972.
4. **Jersild, C., Dupont, B., Fog, T., Platz, P., and Svejgaard, A.**, Histocompatibility determinants in multiple sclerosis, *Transplant. Rev.*, 22, 148, 1975.
5. **Krulig, L., Farber, E.M., Grumet, F.C., and Payne, R.O.**, Histocompatibility (HL-A) antigens in psoriasis, *Arch. Dermatol.*, 111, 857, 1975.
6. **Seignalet, J., Clot, J., Guilhou, J.J., Duntze, F., Meynadier, J., and Robinet-Levy, M.**, HLA antigens and some immunological parameters in psoriasis, *Tissue Antigens*, 4, 59, 1974.

7. **Shreffler, D.C. and David, C.S.**, The H-2 major histocompatibility complex and the I immune response region: genetic variation, function and organization, *Adv. Immunol.*, 20, 125, 1975.
8. **Lobitz, W.**, Summary of the Dermatology Workshop of the HLA and Disease Symposium, Paris, June 1976.
9. **Svejgaard, A.**, personal communication, 1976.
10. **Grosse-Wilde, H., Wustner, B., Albert, E.D., Kuntz, B., Netzel, B., Scholz, S., and Braun-Falco, O.**, Frequency of seven HLA-D alleles in psoriasis patients, in *HLA and Disease*, INSERM, Paris, 1976, 102.
11. **Marcusson, J., Elman, A., Möller, E., and Thyresson, N.**, Psoriasis, sacro-iliitis, and peripheral arthritis occurring in patients with the same HLA haplotype, *Tissue Antigens*, 8, 131, 1976.
12. **Svejgaard, A., Svejgaard, E., Staub Nielsen, L., and Jacobsen, B.**, Some speculations on the associations between HL-A and disease based on studies of psoriasis patients and their families, *Transplant. Proc.*, 5, 1797, 1973.
13. **Tsuji, K., Nose, Y., Ito, M., Ozawa, A., Matsuo, I., Niizuma, K., and Ohkido, M.**, HLA antigens and susceptibility to psoriasis vulgaris in a non-Caucasian population, *Tissue Antigens*, 8, 29, 1976.
14. **Karvonen, J.**, HL-A antigens in psoriasis with special reference to the clinical type, age of onset, exacerbations after respiratory infections and occurrence of arthritis, *Ann. Clin. Res.*, 7, 301, 1975.
15. **Karvonen, J., Tiilikainen, A., and Lassus, A.**, HL-A antigens in patients with persistent palmoplantar pustulosis and pustular psoriasis, *Ann. Clin. Res.*, 7, 112, 1975.
16. **Beckman, L., Bronnestam, R., Cedergren, B., and Liden, S.**, HL-A antigens, blood groups, serum groups, and red cell enzyme types in psoriasis, *Human Hered.*, 24, 496, 1974.
17. **Schoefinius, H.H., Braun-Falco, O., Scholz, S., Steinbauer-Rosenthal, I., Wank, R., and Albert, E.D.**, Histokompatibilitätsantigene (HL-A) bei Psoriasis. Untersuchung an 104 unverwandten Patienten, *Dtsch. Med. Wochenschr.*, 99, 440, 1974.
18. **Woodrow, J.C., Dave, V.K., Usher, N., and Anderson, J.**, The HL-A system and psoriasis, *Br. J. Dermatol.*, 92, 427, 1975.
19. **Svejgaard, A., Staub Nielsen, L., Svejgaard, E., Kissmeyer-Nielsen, F., Hjortshøj, A., and Zachariae, H.**, HL-A in psoriasis vulgaris and in pustular psoriasis-population and family studies, *Br. J. Dermatol.*, 91, 145, 1974.
20. **Katz, S.I., Falchuk, Z.M., Dahl, M.V., Rogentine, G.N., and Strober, W.**, HL-A8: a genetic link between dermatitis herpetiformis and gluten-sensitive enteropathy, *J. Clin. Invest.*, 51, 2977, 1972.
21. **White, A.G., Barnetson, R., DaCosta, J.A., and McClelland, D.B.**, The incidence of HL-A antigens in dermatitis herpetiformis, *Br. J. Dermatol.*, 89, 133, 1973.
22. **Gebhard, R.L., Katz, S.I., Marks, J., Shuster, S., Trapani, R.J., Rogentine, G.N., and Strober, W.**, HL-A antigen type and small-intestine disease in dermatitis herpetiformis, *Lancet*, 2, 760, 1973.
23. **Solheim, B.G., Ek, J., Thune, P.O., Baklein, K., Bratlie, A., Rankin, B., Thoresen, A.B., and Thorsby, E.**, HLA antigens in dermatitis herpetiformis and coeliac disease, *Tissue Antigens*, 7, 57, 1976.
24. **Reunala, T., Salo, O.P., Tiilikainen, A., and Mattila, M.J.**, Histocompatibility antigens in dermatitis herpetiformis with special reference to jejunal abnormalities and acetylator phenotype, *Br. J. Dermatol.*, 94, 139, 1976.
25. **Seah, P.P., Fry, L., Kearney, J.W., Campbell, E., Mowbray, J.F., Stewart, J.S., and Hoffbrand, A.V.**, A comparison of histocompatibility antigens in dermatitis herpetiformis and adult coeliac disease, *Br. J. Dermatol.*, 94, 131, 1976.
26. **Thomsen, M., Platz, P., Marks, J., Ryder, L.P., Shuster, S., Svejgaard, A., and Young, S.H.**, Association of LD-8a and LD-12a with dermatitis herpetiformis, *Tissue Antigens*, 7, 60, 1976.
27. **Roberts-Thomson, I.C., Stevens, D.P., Michel, B., Braun, W.E., Morris, P.J., Wall, A.J., Fone, D.J., and Dworken, H.J.**, Factors influencing small bowel changes in dermatitis herpetiformis, *Aust. N. Z. J. Med.*, 7, 356, 1977.
28. **Falchuk, Z.M., Rogentine, G.N., and Strober, W.**, Predominance of histocompatibility antigen HL-A8 in patients with gluten-sensitive enteropathy, *J. Clin. Invest.*, 51, 1602, 1972.
29. **Strober, W., Nelson, D.L., Rogentine, G.N., and Falchuk, Z.M.**, HL-A antigens and coeliac disease, *Lancet*, 2, 459, 1974.
30. **Keuning, J.J., Pena, A.S., van Hooff, J.P., van Leuwen, A., and van Rood, J.J.**, HLA-Dw3 associated with coeliac disease, *Lancet*, 1, 506, 1976.
31. **Mann, D.L., Nelson, D.L., Katz, S.I., Abelson, L.D., and Strober, W.**, Specific B-cell antigens associated with gluten-sensitive enteropathy and dermatitis herpetiformis, *Lancet*, 1, 110, 1976.
32. **Solheim, B. G., Albrechtsen, D., Thorsby, E., and Thune, P.**, Strong association between an HLA-Dw3 associated B cell alloantigen and dermatisis herpetiformis, *Tissue Antigens*, 10, 114, 1977.
33. **Katz, S.I., Dahl, M.V., Penneys, N., Trapani, R.J., and Rogentine, N.**, HL-A antigens in pemphigus, *Arch. Dermatol.*, 108, 53, 1973.
34. **Krain, L.S.**, Histocompatibility antigens: a laboratory and epidemiologic tool, *J. Invest. Dermatol.*, 62, 67, 1974.

35. **Retornaz, G., Betuel, H., and Thivolet, J.**, HLA antigens in bullous pemphigoid, in *HLA and Disease*, INSERM, Paris, 1976, 115.
36. **Retornaz, G., Betuel, H., Ortonne, J.P., and Thivolet, J.**, HLA antigens and vitiligo, in *HLA and Disease*, INSERM, Paris, 1976, 114.
37. **Saurat, J-H., Cosne, A., Puissant, A., Nunez-Roldan, A., and Hors, J.**, HLA-A and B markers in lichen planus, in *HLA and Disease*, INSERM, Paris, 1976, 117.
38. **Mercier, P. and Zini, G.**, Ehlers-Danlos syndrome and HLA haplotype?, in *HLA and Disease*, INSERM, Paris, 1976, 111.
39. **Laurentaci, G. and Dioguardi, D.**, HLA antigens in keloids and hypertrophic scars, in *HLA and Disease*, INSERM, Paris, 1976, 108.
40. **Kianto, U., Reunala, T., Karvonen, J., Lassus, A., and Tiilikainen, A.**, HLA antigens in alopecia areata, in *HLA and Disease*, INSERM, Paris, 1976, 106.
41. **Kuntz, B., Selzle, D., Braun-Falco, O., Scholz, S., and Albert, E.D.**, HLA-antigens in alopecia areata, in *HLA and Disease*, INSERM, Paris, 1976, 107.
42. **Karvonen, J. and Tiilikainen, A.**, HLA antigens in Hailey-Hailey's disease, *Tissue Antigens*, 8, 277, 1976.
43. **Dick, H.M., Mackie, R., and DeSousa, M.B.**, HLA and mycosis fungoides, in *HLA and Disease*, INSERM, Paris, 1976, 99.
44. **Giraldo, G., Degos, L., Beth, E., Gharbi, R., Day, N., Dastot, H., Hans, M., Reboul, M., and Schmid, M.**, HLA antigens in 16 families with xeroderma pigmentosum, *Tissue Antigens*, 9, 167, 1977.
45. **Challacombe, S.J., Batchelor, R., Kennedy, L., and Lehner, T.**, HLA antigens in recurrent oral ulceration, in *HLA and Disease*, INSERM, 1976, 96.
46. **Ohno, S., Aoki, K., Sugiura, S., Nakayama, E., Itakura, K., and Aizawa, M.**, HL-A5 and Behçet's disease, *Lancet*, 2, 1383, 1973.
47. **Ohno, S., Nakayama, E., Sugiura, S., Itakura, K., Aoki, K., and Aizawa, M.**, Specific histocompatibility antigens associated with Behçet's disease, *Am. J. Ophthalmol.*, 80, 636, 1975.
48. **Rosselet, E., Saudan, Y., and Jeannet, M.**, Recherche des antigènes HL-A dans la maladie de Behçet, *Ophthalmologica*, 172, 116, 1976.
49. **Ersoy, F., Berkel, A.I., Firat, T., and Kazokoglu, H.**, HLA antigens associated with Behçet's disease, in *HLA and Disease*, INSERM, Paris, 1976, 100.
50. **Godeau, P., Torre, D., Campinchi, R., Bloch-Michel, E., Schmid, M., Nunez-Roldan, A., Hors, J., and Dausset, J.**, HLA-B5 and Behçet's disease, in *HLA and Disease*, INSERM, Paris, 1976, 101.
51. **Nakagawa, J., Ikehara, Y., Ito, K., and Fukase, M.**, HLA antigens in various autoimmune and related diseases, in *HLA and Disease*, INSERM, Paris, 1976, 205.
52. **Hoshino, K., Inouye, H., Unokuchi, T., Ito, M., Tamaoki, N., and Tsuji, K.**, HLA and Diseases in Japanese, in *HLA and Disease*, INSERM, Paris, 1976, 249.
53. **Takano, M., Miyajima, T., Kiuchi, M., Ohmori, K., Amemiya, H., Yokoyama, T., Hashizume, H., Iwasaki, Y., Okamoto, S., and Sato, H.**, Behçet's disease and the HLA system, *Tissue Antigens*, 8, 95, 1976.
54. **Chamberlain, M.A.**, Behçet's disease, *Ann. Rheum. Dis.*, 34 (Suppl. 1), 53, 1975.
55. **Nakagawa, J., Ikehara, Y., Ito, K., and Fukase, M.**, HLA antigens in various autoimmune and related diseases, in *HLA and Disease*, INSERM, Paris, 1976, 205.
56. **Krain, L.S. and Terasaki, P.I.**, HLA types in atopic dermatitis, *Lancet*, 1, 1059, 1973.
57. **Goudemand, J., Deffrenne, C., and Desmons, F.**, HLA antigens and atopic dermatitis, in *HLA and Disease*, INSERM, Paris, 1976, 179.
58. **Scholz, S., Ziegler, E., Braun-Falco, O., Wüstner, H., Kuntz, B., and Albert, E.D.**, HLA family studies in patients with atopic dermatitis, in *HLA and Disease*, INSERM, Paris, 1976, 187.
59. **Benedetto, A. V., Braun, W. E., and Taylor, J. S.**, HLA antigens in a family with porphyria cutanea tarda and low uroporphyrinogen decarboxylase, *Clin. Res.*, 26, 5672, 1978.

VI. ENDOCRINOLOGY

A. Juvenile Diabetes Mellitus (JDM)

The first antigen shown to occur with increased frequency in this disease is B8 (combined uncorrected $P < 1.0 \times 10^{-10}$) with a relative risk (RR) of 2.13 and 95% confidence interval of 1.75 to 2.58.[1] The major studies in this disease are shown in Table 9.

TABLE 9

Juvenile Diabetes Mellitus (Primarily Caucasians)

HLA Ag	Controls with Ag (%)	Patients with Ag (%)	Patients with Ag (n)	Patients total (n)	Relative risk	Ref.
B8	25.0	35.7	25	100	1.7	2
	20.0	24.0	12	50	1.3	3
	31.8	49.3	74	150	2.1	6
	23.7	44.7	38	85	2.6	10
	16.3	20.3	12	59	1.3	7
	20.5	30.4	34	112	1.7	33
	11.2	31.1	28	90	3.5	8
	18.0	42.0	28	64	3.5	11
B15	7.0	17.9	5	28	2.9	2
	10.0	36.0	18	50	4.9	3
	12.0	20.7	31	150	1.9	6
	17.9	32.9	28	85	2.3	10
	10.0	15.3	9	59	1.7	7
	15.0	25.0	28	112	1.9	33
	13.2	11.1	10	90	0.8	8
	9.6	23.4	15	64	2.9	11
B7	32.0	28.6	8	28	0.87	2
	21.0	20.0	10	50	0.96	3
	32.2	16.0	8	50	0.42	6
	26.8	10.6	9	85	0.34	10
	15.6	11.9	7	59	0.77	7
	26.0	13.4	15	112	0.45	33
	26.0	14.1	9	64	0.47	11
Dw3	19.1	37.5	24	64	2.5	11

In 1972, the initial study by Finkelstein of insulin-dependent diabetes in 44 patients that included 28 Caucasians, 10 Mexicans, and 6 Negroes concluded that there was no significant difference in any of the HLA antigen frequencies in juvenile diabetes mellitus.[2] Their data showed that 36% of the diabetics had B8 compared to 25% of controls and that 18% had B15 compared to 7% of controls, but they did not show the striking decrease in B7 that later studies cumulatively showed.

In a study of 50 insulin-dependent diabetics in 1973, Singal confirmed the fact that the antigen B15 was significantly increased. Thirty-six percent of his patients compared to 10% of the control population had B15, yielding a relative risk of 4.9.[3] No other studies in diabetes have found such a high relative risk or frequency of this antigen, though the trend toward an increased frequency of B15 has been generally supported.

Ludwig detected a significant decrease in B7, both in his own study and in pooled data.[4] Of a total of 311 patients with juvenile-onset diabetes mellitus, only 41 patients (14.1%) had B7 compared to 26.8% of 3650 controls. The increased frequency of B8 and B15 was not, then, compensated by an evenly distributed decrease among the B-series antigens but seemed to be specifically associated with a decrease in B7. B7 was also observed in only 10 of 68 (14.7%) blood relatives of the diabetic patients. Most remarkable was the fact that B7 could not be detected in a single relative under the age of 35 years with glucose intolerance, whereas in all other blood relatives, the B7 frequency was essentially equal to that in the control group.

The additive effect of B8 and B15 as risks in juvenile diabetes mellitus was noted by Svejgaard,[5] a finding which indicated that such antigen susceptibilities may work through different mechanisms. Interestingly, neither the B8 nor B15 homozygotes had an increased incidence of juvenile diabetes mellitus. Further support for the independent action of B8 and B15 was provided by Nelson, who found B8 increased only in concordant and B15 increased in both concordant and discordant identical twins with diabetes.

Another antigen that is in the cross-reacting group with B15, namely, B18, has been noted to be increased in studies by Cudworth, Seignalet, and Cathelineau[6-8] with frequencies in the patient population of 8 vs. 4.3% of controls, 23.7 vs. 12.2% of controls, and 31.1 vs. 13.8% of controls, respectively. Still another antigen in the B15 cross-reacting group, Bw35, was found to be increased in a study of 32 juvenile diabetics from Sardinia.[9] The frequency of Bw35 was 29.2% compared to 11.4% of 96 controls from the same area, B15 was 12.5% compared to 2.0% in controls, and B8 was similar in the two groups.[9]

The C-series antigen Cw3 and the D-series antigen Dw3 are also increased in JDM.[10] Forty-two patients with juvenile diabetes mellitus were typed with HTC for Dw3, then called LD-8a, and 79 for LD-W15a.[10] Using a relative response of 60% as the cut-off level between a typing response and a nontyping response, among the 42 juvenile diabetics who were typed for Dw3, 21 had Dw3, which was 6 more than the number having B8. All of the B8-positive patients also had the Dw3 specificity. In contrast, of the 79 diabetics tested for LD-W15a, 33 had that antigen. Six individuals who had B15 did not have LD-W15a, though nine had it who were B15-negative. In this study, the relative risk for patients with B8 was 2.4 and increased to 4.5 with Dw3 and was 2.5 with Bw15 and increased to 3.7 with LD-W15a, indicating that both of these D-series antigens had a closer association with juvenile diabetes mellitus than did the B-series antigen. These authors presented preliminary evidence suggesting that those with B8 or B15 and juvenile diabetes mellitus produced a migration inhibitory factor when their lymphocytes were tested against fetal calf pancreas antigen.[10] In another study, Ludwig also found Dw3 increased in JDM (37.5% compared to 19.1% of controls).[11]

Some of the differences noted in these studies may be due to the variable age cut-off for inclusion of patients, for example, less than 30 years by Cudworth[6] and Seignalet,[7] less than 35 years by Cathelineau[8] and Mayr,[12] and less than 40 years by Nerup.[13] B15 and Cw3 were very low, 6 and 12%, respectively, in JDM up to 15 years but rose sharply above that age to 30% and 49%, respectively.[11] Thus, in JDM, there is an increased frequency of B8, B15 (or its cross-reacting antigens B18 and Bw35), Cw3,Dw3, as well as a diminished frequency of B7 that may be a significant protective factor in the development of this disease. A similar lowering of B7 has been noted in celiac disease.

Numerous studies have confirmed the lack of any association with an HLA antigen in maturity-onset diabetes (MOD).[3,6,10,14] But a special subgroup of MOD having pancreatic islet cell antibody (PICA) did show an A1 and B8 association of 56% and 64%, respectively.[15] Also, in two families with MOD, there was evidence for a dominant gene associated with different HLA haplotypes: A3-B15 in 7 of 10 people in one family and Aw19-Bw35 in 9 of 9 members of another family.[16] In 10 families, 23 juvenile diabetic siblings were HLA-identical except for 3 cases. Other HLA-identical siblings thought to be unaffected had abnormal cortisol-primed glucose tolerance tests.[16]

In a study of 30 families in which two or more siblings had juvenile onset diabetes mellitus, Woodrow found a marked divergence in the affected siblings toward having HLA haplotypes in common ($P = 5 \times 10^{-6}$), which suggested a major diabetic gene in the HLA region and appeared to disallow a simple recessive trait.[17] A large study of 54 families showed B8 to be present in 22 (41%) and Bw15 in 23 (43%) of the patients, which was significantly above the control frequencies of 18 and 11%, respectively.[18] Of the first-degree relatives, 16 of 56 were found to have an abnormal glucose tolerance test, 8 of whom (50%) had B15 and 2 (13%) had B8. Four families in this study were informative for joint segregation of an HLA haplotype with an abnormal glucose tolerance test. Two families were compatible with joint segregation and two were not, therefore suggesting that additional genetic and environmental factors could be essential for the development of juvenile diabetes.

Further support for other genetic or environmental factors came from the study of identical twins and is based on the following concept:

> If all identical twins are concordant for diabetes (i.e. if both twins are diabetic), the cause of diabetes may be either genetic or environmental, but if they are sometimes discordant (i.e. if one twin is diabetic but the other is not), then the difference between them must be due to environmental factors.[19]

In these discordant pairs, diabetes cannot be exclusively an inherited disorder. Tattersall found that discordance occurred in 31 of 96 diabetic twin pairs and that surprisingly, discordance was more common in the younger pairs less than 40 years of age even after 3- to 10-year follow-up.[19]

A high frequency of recombination in the families of patients with JDM studied for D-series antigens was reported by Rubenstein.[20] In contrast to the usual recombination frequency for 2 HLA loci of about 1%, there was a 16% recombination frequency between the A and B, and B and D loci in juvenile diabetes, a result that lacks confirmation.[16,17]

A possible link was sought between the HLA system and low-insulin response in 25 nondiabetic, low-insulin responders and 30 insulin-dependent diabetics.[21] No difference was found in the distribution of HLA antigens in the low-insulin responders, which was interpreted as an indication that low-insulin responsiveness would not be involved in the development of juvenile insulin-dependent diabetes and was not an HLA-linked factor. In a study of 67 insulin-dependent juvenile diabetics in whom insulin antibody production was investigated, B8 was detected in 25% of high responders, 55% of moderate responders, and 33% of nonresponders.[22] The difference in B8 antigen frequency between the high responders and the non- and moderate responders was significant ($P < 0.05$). Another study of insulin antibodies, C-peptide, and HLA types was conducted by Ludvigsson in 102 juvenile diabetics with an age of onset of 1 to 16 years and a duration of disease of 1 to 17 years.[23] B8 was found in 38% of the patients, B15 in 31% of patients, and B8 and/or B15 in 59% of the patients with a combined relative risk of 2.1. Patients with B8 and/or B15 did not differ significantly from other patients with regard to C-peptide levels, and more important, no difference was found in the insulin antibody level and the HLA type.

In contrast to the negative or weak HLA association for insulin antibodies, pancreatic islet cell antibodies (PICA) did show interesting correlations. Pancreatic islet cell antibodies that are principally found in insulin-dependent diabetics occur with a frequency of about 65% within 1 year of diagnosis and fall rapidly thereafter.[15] The natural history of such antibodies must be taken into account when studies are compared for such antibody prevalence. Ninety-nine diabetics with PICA showed a frequency for B8 of 61% and for B15 of 12%, whereas 20 diabetics without these antibodies tested within 3 months of the diagnosis showed a frequency for B8 of only 35%, which was not significantly different from controls. Diabetics in whom these antibodies persisted more than 5 years after diagnosis showed a frequency for B8 of 71% (25 of 35). It was in those with PICA that the A1 association with B8 appeared as might have been expected on the basis of linkage disequilibrium. B15 and Cw3 did not appear to show such a correlation with PICA.[15] A trend toward an association of islet cell antibodies and B8 and B15 was found in 139 patients with JDM but lacked significance.[24] Similarly, thyrogastric antibodies were more frequent but not significantly so in those with B8 and B15.

In a study of the clinical variables of JDM such as age of onset, familial incidence, micro- and macroangiopathy, and hyperlipidemia that were studied in 156 patients, a significant association between diabetic retinopathy and B8 (58.3%) was found.[25] Similarly, diabetic microangiopathy in the form of glomerulosclerosis and retinopathy was associated with A1 and B8 in 110 patients.[26] These data seem to be supported by earlier studies demonstrating the striking frequency of retinopathy found in concordant diabetic identical twins[27] and by the increased frequency of B8 and B15 in concordant identical twins with JDM.[14]

In a study from Japan of 172 diabetics whose clinical categorization was not given, no significant association with HLA antigens in diabetes mellitus was found, suggesting a possible racial difference for disease association as has been found in thyrotoxicosis.[28] The absence of B8 in normal Japanese should be reemphasized.

A seasonal clustering of diabetes mellitus has been reported in B8-positive diabetic children in whom 81% of the B8-positive patients had the onset of diabetes between October and February as opposed to only 33% of the B8-negative children.[29] A possible viral etiology, particularly Coxsackie B4, was suspected because the peak frequency of that infection overlaps the same time period. However, studies by Barbosa[30] and by Garovoy[31] have failed to support the seasonal frequency of diabetes in B8-positive individuals. Nevertheless, a viral etiology seems a promising area for further investigation.[6,32]

B. Thyrotoxicosis (Graves' Disease)

The major antigen associated with this disease in Caucasians is B8 (combined uncorrected $P < 1.0 \times 10^{-10}$) with a relative risk of 2.5 and a 95% confidence interval of 2.04 to 3.12.[1] Major studies of this disease are shown in Table 10.

In 1974, Grumet published a series of 62 Caucasian patients with Graves' disease, 46.8% of whom had B8 compared to 21.2% of controls.[34] In 1975, a study from Ireland and Scotland of 60 patients and 60 matched controls failed to show a significantly increased frequency of B8, though 21 of the patients compared to 16 of the controls did have the B8 antigen.[35] Instead, this study showed significant increases in A1 and B13. Later, stronger confirmation of the B8 association with thyrotoxicosis was presented by Whittingham, who found B8 in 42% of 64 Caucasian patients compared to 24% of 700 controls.[36]

These authors were unable to confirm Grumet's observation that the association between B8 and thyroid-antibody-positive thyrotoxic patients was greater than for the thyrotoxic group as a whole. [36] Not all the patients in this study had thyroid antibodies,

TABLE 10

Thyrotoxicosis in Caucasians and Japanese

HLA Ag	Controls with Ag (%)	Patients with Ag (%)	Patients with Ag (n)	Patients total (n)	Relative risk	Ref.
			Caucasians			
B8	21.2	46.8	29	62	3.2	34
	16.3	35.0	35	100	2.8	7
	26.7	35.0	21	60	1.5	35
	24.9	44.4	36	81	2.4	37
	24.0	42.2	27	64	2.3	36
Dw3	16.0	54.0	14	26	6.3	37
			Japanese			
Bw35	20.5	56.8	25	44	5.0	39

which contrasted with the patients reported earlier by Grumet.[34] However, a recent study by Thorsby of 81 thyrotoxic patients showed not only an increased frequency of B8, which was present in 44.4% of patients compared to 24.9% of controls, but again suggested that there was an association with thyroid microsomal antibodies and/or thyroglobulin antibodies because 57% of the antibody-positive patients carried B8 compared to 33% of the antibody-negative patients.[37] Further analysis of the antibody-positive and negative patients revealed that the B8 frequency significantly increased compared to controls only in those who were antibody producers (uncorrected $P < 0.0005$).

Of 26 patients studied for the Dw3 antigen, then called LD-8a, 54% had the antigen compared to 16% of 45 controls.[37] All the Dw3-positive patients were also B8-positive.

A recent study by Ludwig of 97 thyrotoxic patients, 49 with and 48 without endocrine ophthalmopathy, revealed only a slight increase in B8 in the thyrotoxic patients with ophthalmopathy and no difference from controls in those without ophthalmopathy.[38] But the antigen Cw3 that occurred in 18% of controls was found in 28.5% of the ophthalmopathy subgroup compared to only 6.2% of those without ophthalmopathy, a highly significant difference with a P value of < 0.001.

In a study of Japanese patients with thyrotoxicosis, a striking switch in the HLA antigen association was found. Grumet found in studying 44 patients, that the antigen with the most significantly increased frequency was Bw35, then called W5, which occurred in 57% of patients as compared to only 20% of the 83 controls.[39] Furthermore, 56% of the 34 patients with abnormally elevated serum levels of antithyroid microsomal autoantibody had the Bw35 antigen, whereas in 48 controls tested for the same antibody, none of the 7 who were seropositive had the antigen. Notably, the B8 antigen was absent from both the 44 patients and the 83 controls in this Japanese population where a low frequency of B8 has long been established. These authors considered two major possibilities to explain the difference in their findings. The first was that Graves' disease is not the same disease in Caucasians and in Japanese. But the clinical, histopathologic, and serologic abnormalities; response to therapy; and incidence all contradicted that possibility. The second possibility was that B8 and Bw35 each identified distinct genes affecting the susceptibility to Graves' disease. In their original material, Grumet noted that, in the Caucasian patients, Bw35 was increased almost twofold above that of Caucasian controls.[34] Although further studies would be necessary to confirm this, the suggestion is an interesting one in view of the finding that in juvenile diabetes mellitus, B8 and B15 each seemed to contribute a distinct susceptibility to diabetes mellitus and that the B15 antigen and B18 antigen may be more likely to do this in certain races than in others. Their third hypothesis was based on the possibility that the different HLA antigens may have different linked Ia or disease susceptibility antigen(s), though linkage should apply to loci.

C. Subacute Thyroiditis (de Quervain's)

The most significant antigen association found in this disease was Bw35 (combined uncorrected $P < 1.0 \times 10^{-10}$) with a relative risk of 22.2 and a 95% confidence interval of 10.93 to 45.03.[1] A preliminary study from Czechoslovakia of 15 patients with this disease,[40] which is suspected of having a viral etiology, was confirmed by a second separate study of 30 patients in which Bw35 occurred in 53.3% compared to 9.3% of 314 controls ($P < 0.0005$).[41]

D. Hashimoto's Disease

Studies by Mayr of 25 patients,[42] by van Rood of 55 patients,[43] and by Whittingham of 12 patients[44] failed to disclose any significantly increased frequency of an HLA antigen and conflicted with the data by Farid[45] that showed increases in B8, Bw15, and Bw17.

TABLE 11

Idiopathic Addison's Disease

HLA Ag	Controls with Ag (%)	Patients with Ag (%)	Patients with Ag (n)	Patients total (n)	Relative risk	Ref.
B8	23.7	68.8	22	32	6.9	10
	18.0	26.1	6	23	1.5	47

E. Autonomous Adenomas of the Thyroid

In a study comparing the HLA antigens in this disease and in Graves' disease, it was found that in 40 patients with autonomous adenomas of the thyroid, there was no significant association with any HLA antigen.[46]

F. Idiopathic Addison's Disease

The primary antigen deviation in this disease is B8 (combined uncorrected $P = 7.3 \times 10^{-6}$) with a relative risk of 3.88 and a 95% confidence interval of 2.17 to 7.01.[1] The major studies are shown in Table 11.

The initial study of idiopathic Addison's disease by Thomsen showed that B8 occurred in 68.8% of 32 patients compared to 23.7% of 1967 controls.[10] In 30 of these patients studied for the D-series antigen Dw3, then called LD-8a, and for the LD antigen LD-W15a, it was found that 21 of the 30 Addisonian patients had the Dw3 specificity. All of those who had B8 were included in the Dw3-positive group. In contrast, there were only three of the 30 Addisonian patients who had the LD-W15a D-series antigen. In 21 of the 22 B8-positive Addisonian patients, antibodies were found against adrenal cortical antigens. Of the other 10 patients, 5 had the antiadrenal cortical antibody and 1 of these was Dw3-positive. None of the antibody-negative patients had Dw3. The correlation of B8 and Dw3 with such antibodies was significant ($P < 0.006$ and < 0.002, respectively).

However, somewhat conflicting results were obtained by Schernthaner, who could find no statistically significant difference in the frequency of any HLA antigen in 23 patients with idiopathic Addison's disease.[47] Specifically, B8 occurred with only a slightly increased frequency of 26% compared to 18% in 450 controls. Of the patients, 56% showed adrenal cortical antibodies and 65% a migration inhibitory factor. All 6 B8-positive patients had adrenal antibody, whereas only 7 of the 17 B8-negative patients had antiadrenal antibodies. Thus, the study by Schernthaner, which addressed only the B8 antigen question, very clearly demonstrated an increased B8 frequency in those who had antibody to adrenal cortex though it could not confirm the increased frequency of B8 in the disease group as a whole.

G. Adrenocortical Hyperfunction

In this unspecified form of adrenocortical disease, 125 patients were found to have an increased frequency of A1 (41%) and B8 (22%) compared to 28 and 16%, respectively, in 352 controls.[48]

H. Primary Hyperaldosteronism

In a preliminary study of 27 patients with primary hyperaldosteronism, no significant association was found with any single HLA antigen in the group as a whole. However, when the patients were divided into two groups, 22 with tumors and 5 with hyperplasia, it was noted that all five patients with hyperplasia had an A-series antigen of the cross-reacting group (CREG) consisting of antigens A1, A3, or A11, and four had a B-series antigen of the CREG consisting of either B18 or Bw35.[49]

I. Hypoparathyroidism (Pseudo and Pseudo-Pseudo)

In a family with pseudohypoparathyroidism, short metacarpals were inherited with the A3-Bw35 haplotype.[50] However, eight others with short metacarpals or metatarsals, five with pseudo-pseudo hypoparathyroidism, had no such association and only insignificant increases in B8 (43%), B27 (28%), and B18 (28%) compared to 17, 7, and 7%, respectively, for controls.[51]

J. Congenital Adrenal Hyperplasia (21-Hydroxylase Deficiency)

This autosomal recessive inherited disease was studied in 32 families in which affected siblings were always HLA genotypically identical except for a recombinant.[52] A remarkably close linkage was found between HLA (B and D loci) and a gene for 21 hydroxylase activity.

REFERENCES

1. **Ryder, L.P. and Svejgaard, A.,** Associations Between HLA and Disease: Report from the HLA and Disease Registry of Copenhagen, 1976.
2. **Finkelstein, S., Zeller, E., and Walford, R.L.,** No relation between HL-A and juvenile diabetes, *Tissue Antigens,* 2, 74, 1972.
3. **Singal, D.P. and Blajchman, M.A.,** Histocompatibility (HL-A) antigens, lymphocytotoxic antibodies and tissue antibodies in patients with diabetes mellitus, *Diabetes,* 22, 429, 1973.
4. **Ludwig, H., Schernthaner, G., and Mayr, W.R.,** Is HLA-B7 a marker associated with a protective gene in juvenile-onset diabetes mellitus?, *N. Engl. J. Med.,* 294, 1066, 1976.
5. **Svejgaard, A., Platz, P., Ryder, L.P., Staub-Nielsen, L., and Thomsen, M.,** HL-A and disease associations — a survey, *Transplant. Rev.,* 22, 3, 1975.
6. **Cudworth, A.G. and Woodrow, J.C.,** HL-A system and diabetes mellitus, *Diabetes,* 24, 345, 1975.
7. **Seignalet, J., Mirouze, J., Jaffiol, C., Selam, J.L., and Lapinski, H.,** HL-A in Graves' disease and in diabetes mellitus insulin-dependent, *Tissue Antigens,* 6, 272, 1975.
8. **Cathelineau, G., Cathelineau, L., Hors, J., Schmid, M., and Dausset, J.,** Les groupes HLA dans le diabète à début précoce, *Nouv. Presse Med.,* 5, 586, 1976.
9. **Contu, L., Puligheddu, A., Mura, C., and Gabbas, A.,** HLA antigens in Sardinian patients with diabetes mellitus, in *HLA and Disease,* INSERM, Paris, 1976, 124.
10. **Thomsen, M., Platz, P., Andersen, O., Christy, M., Lyngsoe, J., Nerup, J., Rasmussen, K., Ryder, L.P., Staub-Nielsen, L., and Svejgaard, A.,** MLC typing in juvenile diabetes mellitus and idiopathic Addison's disease, *Transplant. Rev.,* 22, 125, 1975.
11. **Ludwig, H., Mayr, W.R., and Schernthaner, G.,** The importance of HLA-B8, Bw15, Cw3 and B7 in juvenile onset diabetes mellitus, in *HLA and Disease,* INSERM, Paris, 1976, 134.
12. **Mayr, W.R., Ludwig, H., Schernthaner, G., and Eibl, M.,** HLA antigens in families of juvenile onset diabetics, in *HLA and Disease,* INSERM, Paris, 1976, 136.
13. **Nerup, J., Platz, P., Andersen, O., Christy, M., Lyngsoe, J., Poulsen, J.E., Ryder, L., Thomsen, M., Staub- Nielsen, L., and Svejgaard, A.,** HL-A antigens and diabetes mellitus, *Lancet,* 2, 864, 1974.
14. **Nelson, P.G., Pyke, D.A., Cudworth, A.G., Woodrow, J.C., and Batchelor, J.R.,** Histocompatibility antigens in diabetic identical twins, *Lancet,* 2, 193, 1975.
15. **Morris, P.J., Vaughan, H., Irvine, W.J., McCallum, C.J., Gray, R.S., Campbell, C.J., Duncan, L.J., and Farquhar, J.W.,** HLA and pancreatic islet cell antibodies in diabetes, *Lancet,* 2, 652, 1976.
16. **Barbosa, J., Noreen, H., Goetz, F.C., and Yunis, E.J.,** The genetic heterogeneity of diabetes: Histocompatibility antigens (HLA) in families with maturity-onset type diabetes of the young (MODY) and juvenile, insulin-dependent diabetes (JID), in *HLA and Disease,* INSERM, Paris, 1976, 121.
17. **Woodrow, J.C. and Cudworth, A.G.,** The sibship study of HLA and diabetes, in *HLA and Diseases,* INSERM, Paris, 1976, 149.
18. **Landgraf, R., Lander, T., Landgraf-Leurs, M.M.C., Scholz, S., and Albert, E.D.,** HLA-haplotypes and glucose tolerance in families of patients with juvenile onset diabetes mellitus (JOD), in *HLA and Disease,* INSERM, Paris, 1976, 131.
19. **Tattersall, R.B. and Pyke, D.A.,** Diabetes in identical twins, *Lancet,* 2, 1120, 1972.

20. **Rubinstein, P., Suciu-Foca, N., Nicholson, J.F., Fotino, M., Molinaro, A., Harisiadis, L., Hardy, M.A., Reemtsma, K., and Allen, F.H., Jr.**, The HLA system in the families of patients with juvenile diabetes mellitus, *J. Exp. Med.*, 143, 1277, 1976.
21. **Vialettes, B., Vague, Ph., and Mercier, P.**, HLA differences between insulin dependent diabetics and non diabetic low insulin responders, in *HLA and Disease*, INSERM, Paris, 1976, 147.
22. **Schernthaner, G., Ludwig, H., Mayr, W.R., and Willvonseder, R.**, Humoral antiinsulin immunity and HLA factors in juvenile onset diabetes, in *HLA and Disease*, INSERM, Paris, 1976, 143.
23. **Ludvigsson, J., Heding, L., and Säfwenberg, J.**, HLA-types, C-peptide and insulin antibodies in juvenile diabetics, in *HLA and Disease*, INSERM, Paris, 1976, 133.
24. **Lendrum, R., Walker, G., Cudworth, A.G., Woodrow, J.C., and Gamble, D.R.**, HLA-linked genes and islet-cell antibodies in diabetes mellitus, *Br. Med. J.*, 1, 1565, 1976.
25. **Bertrams, J., Reis, H.E., Jansen, F.K., Grüneklee, D., Drost, H., Beyer, J., and Gries, F.A.**, Diabetic retinopathy and HLA antigen B8, in *HLA and Disease*, INSERM, Paris, 1976, 294.
26. **Barbosa, J., Noreen, H., Emme, L., Goetz, F., Simmons, R., de Leiva, A., Najarian, J., and Yunis, E.J.**, Histocompatibility (HLA) antigens and diabetic microangiopathy, *Tissue Antigens*, 7, 233, 1976.
27. **Pyke, D.A. and Tattersall, R.B.**, Diabetic retinopathy in identical twins, *Diabetes*, 22, 613, 1973.
28. **Hoshino, K., Inouye, H., Unokuchi, T., Ito, M., Tamaoki, N., and Tsuji, K.**, HLA and diseases in Japanese, in *HLA and Disease*, INSERM, Paris, 1976, 249.
29. **Rolles, C.J., Rayner, P.H., and Macintosh, P.**, Aetiology of juvenile diabetes, *Lancet*, 2, 230, 1975.
30. **Barbosa, J., Noreen, H., Goetz, F., Simmons, R., Najarian, J., and Yunis, E.J.**, Juvenile diabetes and viruses, *Lancet*, 1, 371, 1976.
31. **Garovoy, M.R., Carpenter, C.B., Myrberg, S.M., Gleason, R.E., Funk, I.B., and Craighead, J.E.**, Increased incidence of HLA-B8 in juvenile onset diabetes mellitus, *Transplant. Proc.*, (Suppl. 1), 177, 1977.
32. **Maugh, T.H.**, Diabetes: Epidemiology suggests a viral connection, *Science*, 188, 347, 1975.
33. **Bertrams, J., Jansen, F.K., Grüneklee, D., Reis, H.E., Drost, H., Beyer, J., Gries, F.A., and Kuwert, E.**, HLA antigens and immunoresponsiveness to insulin in insulin-dependent diabetes mellitus, *Tissue Antigens*, 8, 13, 1976.
34. **Grumet, F.C., Payne, R.O., Konishi, J., and Kriss, J.P.**, HL-A antigens as markers for disease susceptibility and autoimmunity in Graves' disease, *J. Clin. Endocrinol. Metab.*, 39, 1115, 1974.
35. **Nelson, S.D. and Pollet, J.E.**, HL-A antigens and thyrotoxicosis, *Tissue Antigens*, 5, 38, 1975.
36. **Whittingham, S., Morris, P.J., and Martin, F.I.**, HL-A8: A genetic link with thyrotoxicosis, *Tissue Antigens*, 6, 23, 1975.
37. **Thorsby, E., Segaard, E., Solem, J.H., and Kornstad, L.**, The frequency of major histocompatibility antigens (SD & LD) in thyrotoxicosis, *Tissue Antigens*, 6, 54, 1975.
38. **Ludwig, H., Schernthaner, G., Mayr, W.R., and Mehdi, S.Q.**, Increased susceptibility to endocrine ophthalmopathy in HLA CW3 positive thyrotoxicosis patients, in *HLA and Disease*, INSERM, Paris, 1976, 135.
39. **Grumet, F.C., Payne, R.O., Konishi, J., Mori, T., and Kriss, J.P.**, HL-A antigens in Japanese patients with Graves' disease, *Tissue Antigens*, 6, 347, 1975.
40. **Nyulassy, S., Hnilica, P., and Stefanovic, J.**, The HL-A system and subacute thyroiditis. A preliminary report, *Tissue Antigens*, 6, 105, 1975.
41. **Buc, M., Nyulassy, S., Hnilica, P., and Stefanovic, J.**, HLA-Bw35 and subacute (de Quervain's) thyroiditis. A definitive report, in *HLA and Disease*, INSERM, Paris, 1976, 122.
42. **Mayr, W.R., Schernthaner, G., Ludwig, H., Mehdi, S.Q., and Höfer, R.**, Missing evidence of a correlation between Hashimoto-thyroiditis and HLA antigens, in *HLA and Disease*, INSERM, Paris, 1976, 137.
43. **van Rood, J.J., van Hooff, J.P., and Keuning, J.J.**, Disease predisposition, immune responsiveness, and the fine structure of the HL-A supergene, *Transplant. Rev.*, 22, 75, 1975.
44. **Whittingham, S., Youngchaiyud, U., Mackay, I.R., Buckley, J.D., and Morris, P.J.**, Thyrogastric autoimmune disease, *Clin. Exp. Immunol.*, 19, 289, 1975.
45. **Farid, N.R., Barnard, J., Kutas, C., Noel, E.P., and Marshall, W.H.**, HL-A antigens in Graves' disease and Hashimoto's thyroiditis, *Int. Arch. Allergy Appl. Immunol.*, 49, 837, 1975.
46. **Weise, W. and Wenzel, K.-W.**, Distribution of HLA-antigens in Graves' disease and in autonomous adenoma of the thyroid, in *HLA and Disease*, INSERM, Paris, 1976, 148.
47. **Schernthaner, G., Mayr, W.R., and Ludwig, H.**, HLA-antigens and autoimmunity in idiopathic Addison's disease, in *HLA and Disease*, INSERM, Paris, 1976, 144.
48. **Lada, G., Gyodi, E., and Glaz, E.**, HLA antigens in patients with adreno-cortical hyperfunction, in *HLA and Disease*, INSERM, Paris, 1976, 130.
49. **Braun, W.E., Bravo, E.L., and Frank, S.A.**, unpublished observations.
50. **Farriaux, J.P., Delmas, Y., Ropartz, C., and Fontaine, G.**, Brachymetacarpie-Groupe HLA. Liaison dans une famille de pseudo-hypoparathyroidie, *Nouv. Presse Med.*, 4, 589, 1975.

51. **Saint-Hiller, Y., Dupond, J.-L., Perol, C., Herve, P., and Betnel, H.,** Les groupes HLA chez les malades ayant une brachymetacarpie, *Nouv. Presse Med.*, 5, 1003, 1976.
52. **Yang, S. Y., Levine, L. S., Oberfield, S. E., O'Niell, G. J., and DuPont, B.,** Mapping of the 21-hydroxylase deficiency gene within the HLA complex, *Transplant. Proc.*, in press.

VII. GASTROENTEROLOGY

A. Chronic Active Hepatitis (CAH)

The primary antigen shown to occur with increased frequency in this disease is B8 (combined uncorrected $P< 1.0 \times 10^{-10}$) with a relative risk (RR) of 3.04 and 95% confidence interval of 2.25 to 4.13.[1] The major studies in this disease are shown in Table 12.

It was this gastroenterologic disease which was one of the earliest of any disease to have a significant association found with an HLA antigen, in this case, HLA-B8. The initial study by Mackay in 1972 concerned 37 patients with chronic active hepatitis (CAH), 25 of whom (68%) had the antigen B8.[2] This was significantly greater than the 25% frequency found for this antigen in 28 patients with other forms of liver disease and the 18% found in 350 healthy Caucasian controls ($P < 0.0001$ uncorrected.)

TABLE 12

Liver Diseases

HLA Ag	Controls with Ag (%)	Patients with Ag (%)	Patients with Ag (n)	Patients total (n)	Relative risk	Ref.
		Chronic Active Hepatitis				
B8	18.0	67.6	25	37	9.2	2
	23.2	46.3	19	41	2.9	9
	18.0	53.8	14	26	5.3	12
	—	86.0	115	134	25.0	3
(HB_sAg-)	21.0	63.0	31	49	6.4	8
(HB_sAg+)	21.0	0.0	0	8	0.5	8
Dw3	17.0	66.0	25	38	9.4	5
		Healthy Carriers of HB_sAg				
Bw41	1.2	11.9	5	42	11.1	16
		Primary Biliary Cirrhosis				
B15	26.1	52.9	9	17	3.2	12
B15 Orcein +	26.1	66.7	8	12	5.7	12
Bw35 Orcein −	26.1	80.0	4	5	11.4	12
		Idiopathic Hemochromatosis				
A3	31	85.0	17	20	12.2	24
	—	91.7	11	12	—	25
	30.5	74.4	29	39	6.6	22
	27.0	75.0	61	84	7.2	23

In examining the clinical characteristics of CAH, these authors noted that in the 19 patients whose disease onset was between the ages of 14 and 30 years, 12 of whom had positive LE cell tests, 13 had the B8 antigen in conjunction with its companion first series antigen, A1. In 16 patients with adult-onset CAH, 8 of whom had positive LE cell tests, 7 had B8, again in conjunction with A1. The high frequency of B8 seemed to have particular relevance in this disease because of the reported association of B8 with systemic lupus erythematosus. However, this association is not a strong one (see Connective Tissue Diseases [Chapter 7, Section IV]). Nevertheless, B8 has emerged in a variety of other diseases with an immunologic component such as celiac disease, myasthenia gravis, Sjögren's syndrome, and thyrotoxicosis. The association of B8 with CAH has been confirmed in the other studies noted in Table 12, though none has shown as high a relative risk as the initial study of Mackay. The higher frequency of A1 in this disease is believed to be due to the linkage disequilibrium between A1 and B8 and does not seem to have an independent contribution to the relative risk.

In an enlarged study that confirmed their earlier results, Freudenberg studied 134 adult patients and the families of 30 children with CAH.[3] The frequency of B8 in this series increased to 86%.

In a family with 5 SD-identical siblings, all A1, A3, B7, and B8, one of whom had classical CAH, it was found that the maternally inherited B8 antigen was not solely responsible for the genetic predisposition to this disease.[4] The fact that in mixed lymphocyte cultures the patient's lymphocytes responded to and stimulated those of her four siblings, whereas the other four siblings were all mutually unresponsive, indicated a different D-locus antigen in the patient. This finding of at least one other genetic factor being necessary for predisposition to an HLA-associated disease is a recurrent finding in all of the diseases now being reported.

Another investigation of D-series antigens by Opelz and Terasaki demonstrated that the Dw3 antigen found in 24% of 91 normal controls increased to 68% of 38 patients with CAH.[5] These authors found that the association of Dw3 with CAH was stronger ($P < 0.0001$) than that with B8 ($P < 0.001$) and related to a poor therapeutic response to steroids.

A different aspect of the CAH picture was examined by Eddleston,[6] who found that B8 was present in 63% of 57 patients with HB_sAg-negative chronic hepatitis, a frequency significantly higher than the 25% found in 95 control subjects. A more remarkable part of their study was the fact that B8 was not present in any of their eight patients who were HB_sAg-positive with CAH, a level significantly lower than that of the control population.

Other studies have also been directed at the relationship of HLA phenotypes and the persistence of HB_sAg. The frequency of the A1-B8 phenotype was examined in 440 hemodialyzed patients, some of whom had been reported earlier.[7] Serial studies of the HB_sAg categorized 185 of these patients as constantly negative, 185 persistently positive, and 70 transiently positive.[7] Antigens A1-B8, present in the normal French population in 5.05%, were unusually frequent (18.6%) in the 70 patients who eliminated the HB_sAg antigen within a few months. The high frequency of the A1-B8 phenotype in the positive-to-negative seroconvertors paralleled the abnormally high frequency of the same phenotype in patients who had seronegative CAH and suggested that once again, this phenotype might be associated with an augmented immune response against the HB_sAg. Within the group of 185 patients unable to eliminate the HB_sAg antigen, the frequency of the 1-8 phenotype was 7%, essentially that of the control population.

CAH was studied further in terms of antibodies to rubella, measles, smooth muscle, nuclei, and *E. coli* in 57 patients with CAH, 8 of whom were HB_sAg-positive.[8] In those patients with B8 who were HB_sAg-negative, the rubella antibody titers were significantly higher, whereas in those with B12, the antinuclear antibody titer was higher with the suggestion that there was a synergistic effect when B8 and B12 were present together because all seven patients with this combination had significantly higher titers of all the antibodies tested, with the exception of those against *E. coli*. The Hb_sAg-positive patients, all lacking B8, had significantly lower antibody titers than did the HB_sAg-negative cases. Thus, HB_sAg negativity in CAH was found in conjunction with B8 and high antibody titers, a response augmented further by the presence of the B12 antigen. This type of finding showing different degrees of antibody response to viruses in association with certain HLA antigens has obvious potential significance in other infectious diseases.

Lindberg studied 46 patients with CAH of three types: virus-induced, drug-induced, and cryptogenic. An increase in B8 occurred in the drug-induced and cryptogenic forms (41.7 and 62.5%) and was significant in the latter group ($P<0.01$). High titers of smooth muscle and antinuclear antibodies also were found most frequently (about one third to two thirds of the cases) in the drug-induced and cryptogenic forms.[9] These data were interpreted in light of Eddleston's hypothesis that in the HB_sAg-negative patients, who are typically B8-positive, defective suppressor T cell function permits B cells to make liver autoantibodies presumably to cell-membrane lipoprotein.[10] This concept has been supported by recent studies showing a linear pattern of IgG deposition on the hepatocyte basement membrane in those patients who are HB_sAg-negative, a mechanism analogous to the formation of antiglomerular basement membrane antibodies in Goodpasture's syndrome.[11] The granular deposition of IgG found in HB_sAg-positive patients who lack B8 pointed to liver damage by immune complexes.

Cell mediated immunity was investigated in 22 Finnish patients with CAH and a 54% frequency of B8 who had their lymphocytes stimulated by PHA, Con-A, pokeweed mitogen, PPD, and lipopolysaccharide in mixed lymphocyte culture reactions.[12] The responses to PPD and Con-A were similar to those in normal controls, whereas the response to PHA and to pooled human lymphocytes was enhanced. The opposite of this reaction was found in primary biliary cirrhosis. Despite the associations between these types of liver disease and lymphocyte responsiveness, there was no association found between the altered responses and the HLA antigens of the patients.

In Japanese, a switch in the antigen association was noted by Hoshino, who found that in 168 patients with hepatitis, the most frequent antigens were B5 and A29.[13]

B. Asymptomatic Carriers of HB_sAg

The primary antigen shown to occur with increased frequency in this disease (Table 12) may be Bw41 (uncorrected $P< 7.2 \times 10^{-4}$) with a relative risk (RR) of 11.2.[1] However, no consistent alterations in HLA frequency have been found in this clinical state. Vermylen reported an increase of A3 and Aw19 in 48 healthy carriers,[14] Gyodi a decrease in Bw17 and B27 in 83 healthy carriers,[15] Jeannet an increase of Bw41 in 96 individuals,[16] and Seignalet an increase in an antigen cross-reacting with A11 but never more fully defined.[17] Interestingly, except for one group (Gyodi), no decrease in B8 was found. Furthermore, unpublished data of Boettcher revealed a decreased frequency of Bw15 in Australian aborigines with HB_sAg.[18]

C. Primary Biliary Cirrhosis

The primary antigen shown to occur with increased frequency in this disease appears to be B15 with a relative risk (RR) of 3.2[12] A major study in this disease is shown in Table 12.

Seventeen patients with primary biliary cirrhosis were reported from Finland.[12] Of the 17 patients, 9 (53%) had B15 as compared to 26% of the 326 controls, and only 1 of 17 (6%) had B8. In primary biliary cirrhosis, the response of lymphocytes from 12 of these patients to pokeweed mitogen, pooled human lymphocytes, and especially lipopolysaccharide was markedly decreased. When the patients with primary biliary cirrhosis were divided into those who were orcein-positive and orcein-negative, it was found that the association with B15 was most strongly evident in the orcein-positive group of 12 patients, 8 of whom (67%) had the antigen. Orcein is an intracellular copper-binding protein found in liver biopsies from these patients and presumably indicates more advanced cholestatic liver disease. Only one of five (20%) of the orcein-negative patients had the B15. In the same group of patients, another cross-reacting antigen, Bw35, was found in seven patients (41%) as compared to 19% in the controls. This antigen was found in only three (25%) of the orcein-positive group and four (80%) of the orcein-negative group. Thus, two cross-reacting antigens seemed to account for all of the primary biliary cirrhotic patients and even helped to distinguish the orcein-positive from the orcein-negative subgroups.

An earlier study by Galbraith describing 45 patients from southeast England and 28 patients from western Scotland showed a higher frequency of the antigen Bw35 (18% compared to 8% controls) in the latter group but a lower frequency of B15 (4% compared to 8% controls) and no increase in B8.[19]

D. Alcoholic Liver Disease

In 63 patients with alcoholic liver disease, the prevalence of B8 was increased to 45% in those with cirrhosis compared to a frequency of 25% in controls and only 10% in those with a fatty liver or minimal fibrosis.[20] A28 was completely lacking in the cirrhotic patients, and interestingly, was also absent in patients with CAH, suggesting a protective function.

E. Gilbert's Disease

This disease is characterized by an unconjugated hyperbilirubinemia with a normal conjugated serum bilirubin level in the absence of overt hemolysis and pathologic liver function and structure. Its familial incidence raised the suspicion that it might have an association with an HLA antigen. A study by Mayr of 19 unrelated patients with Gilbert's syndrome and 21 first-degree relatives disclosed only a slight and insignificant increase of the antigens A11 and Bw35 in the patients.[21] There was no segregation of the disease with HLA haplotypes within the families. Thus far, there appears to be no association between the HLA system and any hypothetical locus governing Gilbert's syndrome.

F. Idiopathic Hemochromatosis

The primary antigen shown to occur with increased frequency in this disease is A3 with a relative risk (RR) of 9.49.[1] The major studies in this disease are shown in Table 12.

Several very recent studies have confirmed a significant increase in A3 in idiopathic hemochromatosis. In 39 cases from England, the frequency of A3 was 74.4% compared to 30.5% in 95 normal controls.[22] The largest series comprised 84 unrelated patients and was reported by Fauchet, who established the diagnosis with the following criteria: skin pigmentation, hepatomegaly, diabetes mellitus, gonadal deficiency, heart failure, high plasma iron level and a low unsaturated iron-binding capacity, abnormal urinary iron excretion by the desferrioxiamine test, and the demonstration of iron overload by liver biopsy. [23] Among the 84 patients, HLA-A3 was found in 75% compared to 27% of 204 healthy subjects. In addition, the second-series antigen B14 was found in 31% compared to 3.4% of controls. These authors calculated the relative risk for the individual antigens, A3 and B14, as well as for the A3-B14 combination and found them to be 8.2, 26.7, and 90, respectively. Five families from this study showed an association of the disease with the haplotype A3-B14. The two studies of idiopathic

TABLE 13

Gluten Sensitive Enteropathy

HLA Ag	Controls with Ag (%)	Patients with Ag (%)	Patients with Ag (n)	Patients total (n)	Relative risk	Ref.
B8	29.5	87.7	43	49	17.2	30
	21.5	87.5	21	24	22.2	29
	19.0	66.0	35	53	8.1	37
	27.3	76.2	16	21	8.0	31
	29.5	60.0	18	30	3.6	38
	16.3	55.6	30	54	6.3	39
	11.2	45.3	24	53	6.4	40
	22.0	82.5	33	40	15.8	41
	24.9	67.4	31	46	6.1	33
	26.1	88.9	32	36	20.3	42
	35.3	75.0	30	40	5.5	43
	17.0	51.0	27	53	5.1	44
	18.0	70.0	14	20	10.6	45

hemochromatosis by Simon[24] and Shewan[25] that first showed increases in A3 are thus amply confirmed.

G. Chronic Pancreatitis

In an unconfirmed preliminary report by Gastard, 20 unrelated patients of Breton descent who had chronic primitive pancreatitis were found to have A1 in 55% (11 patients) compared to 23.7% in the control group, Bw40 in 30% (6 patients) compared to 3.6% in the control group, and B8 in only 1 patient despite the known linkage disequilibrium of A1-B8.[26] The suggestion from this study that A1 and B40 are increased and B8 decreased in chronic primitive pancreatitis will require confirmation.

H. Cystic Fibrosis (CF)

Twenty-four patients from 5 months to 16 years of age had no significant alteration in HLA antigen frequency, though B5 was increased to 29% from 14% in controls.[27] However, a study of 12 homozygous patients and 32 heterozygous gene carriers in families with CF showed an increase in B18 to 50 and 31%, respectively, compared to 14% in the normal population.[28]

Therefore, in CF, there is a tendency for an increase in the frequency of B5-B18 cross-reacting antigens similar to what has been reported in other diseases such as Hodgkin's.

I. Gluten Sensitive Enteropathy (GSE)

The first antigen shown to occur with increased frequency in this disease is B8 (combined uncorrected $P < 1.0 \times 10^{-10}$) with a relative risk (RR) of 8.84 and 95% confidence interval of 6.98 to 11.18.[1] The major studies in this disease are shown in Table 13.

A large number of studies have now confirmed the high frequency of B8 in patients with gluten sensitive enteropathy. In approximately 800 patients with GSE who have now been studied, B8 has been found to incur a relative risk of about 8 to 10 times. B8 was found in 60 to 65% of children and in 80 to 85% of adults with GSE, suggesting a relationship of this antigen to the age of onset of this disease. In families, B8 segregated with the disease but was not the only antigen associated with it. The decrease noted in B7 may reflect the fact that in general, persons with B7 have a decreased immune response or that B7 confers a relative resistance to celiac disease.

Further details of the HLA and GSE association are summarized below. In 1972, two studies of 24[29] and 49[30] GSE patients showed that 87.5% (21/24) of the patients in the first study had B8 and 88% (43/49) in the second study. In the Falchuk study, three families of patients with B8 and GSE showed an autosomal dominant type of inheritance.[29] These authors speculated that GSE, which is clearly related to the ingestion of a food protein, gluten, may have an Ir gene which was linked to the B locus and which controlled the immunologic manifestations in this disease such as the increased serum IgA levels and increased mucosal synthesis of IgA in large part due to the synthesis of antigluten antibodies. An alternate possibility was also considered, namely, that B8 may have represented a surface receptor site where an antigen such as gluten could bind and lead to local antibody production and tissue injury. Even at this early stage, these authors pointed out that B8 could not be the sole factor responsible for the occurrence of GSE because not every patient with GSE had B8, at least 12% being negative, and because the 20% of the normal population having B8 did not suffer from the disease. Furthermore, a family study indicated that HLA-identical siblings, both with B8, had a discordant occurrence of gluten enteropathy. The increased frequency of A1 was most probably related to its linkage disequilibrium with B8 in both of these studies (66.7% in Falchuk's study[29] and 78% in Stokes' study).[30]

Commenting on the status of the GSE studies at the end of 1973, Evans observed that the association of B8 with celiac disease was highly significant in five reports, that persons with B8 had approximately ten times the risk of having this disorder compared with persons possessing other antigens, and finally, that there was no significant heterogeneity between the series published from geographically different areas.[31]

The next major advance in the study of gluten sensitive enteropathy came from van Rood's group, who found that the D-series antigen Dw3 was more significantly associated with GSE than was B8.[32] In this study, 22 of 28 patients had B8, and 27 of 28 had Dw3. Despite the remarkably high association of Dw3 with GSE, family studies by these same authors indicated a polygenic inheritance of GSE, since there were relatives of these patients who possessed Dw3 but who had no indication of the disease either clinically or morphologically. One of the more interesting aspects of this study was the fact that in addition to determining the Dw3 specificity by homozygous LD typing cells (HTC), these authors were able to use two serologic reagents, sera Mo and Be, which also appeared to recognize the Dw3 determinant with the same high level of significance. These findings of a closer association with the D-series antigens than with the B-series antigens places GSE in the same category as multiple sclerosis with the Dw2 antigen, rheumatoid arthritis with the Dw4 antigen, and others.

However, in children with GSE, Solheim reported a comparable incidence of B8 and Dw3.[33] B8 was present in 67.4% of the 46 children tested compared to 24.9% of 1628 unrelated controls, and similarly, Dw3 was found in 61.5% of 26 patients tested compared to 19% of 136 unrelated controls.

An approach similar to van Rood's for serologically defining the antigens associated with GSE was taken by Mann, who found that two antisera, one derived from the spouse (W-1) and the other from the mother (B-1) of two patients with GSE, reacted with the enriched B-cell preparations of 15 of 16, and 13 of 16 GSE patients, respectively.[34] These antigens did not segregate with HLA, though other B-cell antigens reported by the same author were linked to HLA in studies of large Amish families.[35] The antigens detected by these antisera were thought to be possibly human analogues of the murine I region antigens. [35] Such an antigen might represent the gene product of an abnormal Ir gene in man which could produce an immune response to wheat protein and induce immunologically mediated injury to the gut epithelium. Another possibility which overlaps with the first is that the B-lymphocyte antigen thus defined might be a receptor or binding site for wheat protein and thereby lead to immune reactions in the gut epithelium. Evidence along this line was presented by Nelson,[36] who studied the alkaline phosphatase activity in jejunal biopsies at the initiation of and after 24 to 48 hr of culture in the presence or absence of alpha gliadin and a fragment of alpha gliadin. They found that when biopsies were cultured in the presence of alpha gliadin, the normal rise in alkaline phosphatase activity was inhibited in GSE patients who possessed B8, whereas no inhibition was seen in patients lacking B8. However, when the alpha gliadin fragment was tested, inhibition of the rise in alkaline phosphatase activity was found in all patients regardless of their HLA antigen status. An hypothesis offered was that GSE patients may have a binding site for gluten proteins on the epithelial cells and/or lymphocytes, which depends on the HLA gene product B8, with the result that B8-positive individuals may react to alpha gliadin or a subunit of alpha gliadin, whereas B8-negative individuals react only to the subunit of alpha gliadin.

An association of B8 and Dw3 with dermatitis herpetiformis has also been described and is discussed under Dermatology (Chapter 7, Section V).[29,32-34]

J. Inflammatory Bowel Disease (IBD)

As noted under Rheumatology (Section VII.N), the inflammatory bowel diseases of ulcerative colitis and Crohn's disease had no specific HLA antigen increase by themselves. Studies by Mallas of 100 patients with Crohn's disease and 100 patients with ulcerative colitis confirmed the lack of association of HLA and these diseases.[46] Murtagh studied 30 patients of Irish descent with ulcerative colitis and showed no significant HLA association.[43] In 42 French patients with ulcerative colitis and 24 with Crohn's disease, Vachon showed no significant deviation of antigen frequency.[47]

A raised frequency of HLA antigens in IBD was suggested in reports from the Netherlands,[48] Sweden,[49] and Japan.[50] Vandenberg Loonen found an increased frequency of A11 in 58 patients with ulcerative colitis, whereas there was an increased frequency of B18 in 51 patients with Crohn's disease.[48] In 62 patients with Crohn's disease, an insignificant increase in B17 was found which, when combined with data from two early studies, showed an increased relative risk of 3.24.[49] No difference was found in ulcerative colitis.[49] In 40 Japanese patients with ulcerative colitis, Tsuchiya found an elevation of B5 and of the haplotype A9-B5.[50] However, with the exception of Bergman's results,[49] no confirmatory studies exist for these findings, and in fact, most studies have had negative results.[1]

K. Acute Appendicitis

In a rather surprising finding from Bulgaria, the antigen B12 was found to be elevated to 31% among 696 patients with histologically verified acute appendicitis, as compared to 18.3% in 1085 controls.[51] The authors suggested that an increased frequency of B12 might help to differentiate appendicitis from other acute disorders of the digestive tract and that its lymphoid composition might be relevant to the B12 association.

L. Pyloric Stenosis

No significant alteration in HLA antigen frequency as well as no confirmation of a decrease in blood group A were reported in 105 children who had had surgery for infantile pyloric stenosis.[52]

M. Pernicious Anemia

The primary antigen shown to occur with increased frequency in this disease (Table 14) is B7 (combined uncorrected $P < 1.2 \times 10^{-4}$) with a relative risk (RR) of 2.20 and 95% confidence interval of 1.47 to 3.30.[1] The three studies in Table 14 have shown a

TABLE 14

Pernicious Anemia

HLA Ag	Controls with Ag (%)	Patients with Ag (%)	Patients with Ag (n)	Patients total (n)	Relative risk	Ref.
B7	26.8	51.7	15	29	2.9	54
	19.0	38.1	16	42	2.6	55
	22.1	27.3	18	66	1.3	53

small but increased frequency of B7 in a total of 137 patients with pernicious anemia.[53-55] As described under Neurology (Chapter 7, Section X), when patients with the neurologic complications of pernicious anemia were tested, Horton found an increased frequency of the phenotype A2-B12 (58% compared to only 6% of 54 patients with pernicious anemia without neurologic damage, P<0.00025).[53]

When tested, 46[54] and 27 [56] patients with autoimmune atrophic gastritis showed an insignificant increase in A3. No association was found with gastric parietal cell or intrinsic factor antibody. An explanation for the increased B7 in pernicious anemia but not in atrophic gastritis when both have the same histology might be that the typical progression to pernicious anemia is influenced by HLA. Therefore, this finding veers away from the tendency of the endocrinologic disorders such as Addison's disease and juvenile diabetes mellitus to show a closer association with an HLA antigen when that disease entity is looked at in terms of autoantibody production.

N. Acetylation Capacity

The rate of drug acetylation by the liver was measured in 50 patients with dermatitis herpetiformis. No association was found between the rate of acetylation and any HLA antigen, even B8.[57]

O. Dermatitis Herpetiformis (see Dermatology [Chapter 7, Section V])

REFERENCES

1. **Ryder, L.P. and Svejgaard, A.,** Associations Between HLA and Disease: Report from the HLA and Disease Registry of Copenhagen, 1976.
2. **Mackay, I.R. and Morris, P.J.,** Association of autoimmunne active chronic hepatitis with HL-A1, 8, *Lancet,* 2, 793, 1972.
3. **Freudenberg, J., Baumann, H., Arnold, W., and Meyer Zum Buschenfelde, K.H.,** HLA in different forms of chronic active hepatitis. A comparison between children and adult patients, in *HLA and Disease,* INSERM, Paris, 1976, 158.
4. **Dumble, L.J. and Mackay, I.R.,** HLA and chronic active hepatitis (CAH), in *HLA and Disease,* INSERM, Paris, 1976, 154.
5. **Opelz, G., Vogten, A., Summerskill, W., Terasaki, P., and Schalm, S.,** HLA determinants in chronic active liver disease: Possible relation of HLA-Dw3 to prognosis, *Tissue Antigens,* 9, 36, 1977.
6. **Eddleston, A.L.W.F., Galbraith, R.M., Williams, R., Kennedy, L.A., and Batchelor, J.R.,** HLA-B8 and chronic active hepatitis, in *HLA and Disease,* INSERM, Paris, 1976, 155.
7. **Jungers, P., Descamps, B., Naret, C., Zingraff, J., and Bach, J.F.,** HLA-A1, B8 phenotype and HBs antigenemia evolution in 440 hemodialyzed patients, in *HLA and Disease,* INSERM, Paris, 1976, 160.
8. **Galbraith, R.M., Williams, R., Pattison, J., Kennedy, L.A., Eddleston, A.L., Webster, A.D., Doniach, D., and Batchelor, J.R.,** Enhanced antibody responses in active chronic hepatitis: relation to HLA-B8 and HLA-B12 and porto-systemic shunting, *Lancet,* 1, 930, 1976.

9. **Lindberg, J., Lindholm, A., Lundin, P., and Ivarson, S.**, Trigger factors and HL-A antigens in chronic active hepatitis, *Br. Med. J.*, 4, 77, 1975.
10. **Eddleston, A.L. and Williams, R.**, Inadequate antibody response to HB Ag or suppressor T-cell defect in development of active chronic hepatitis, *Lancet*, 2, 1543, 1974.
11. **Paronetto, F. and Popper, H.**, Two immunologic reactions in the pathogenesis of hepatitis?, *N. Engl. J. Med.*, 294, 606, 1976.
12. **Salaspuro, M., Makkonen, H., Sipponen, P., and Tiilikainen, A.**, HLA-B8, HLA-Bw15 and lymphocyte stimulation in chronic active hepatitis and primary biliary cirrhosis, in *HLA and Disease*, INSERM, Paris, 1976, 168.
13. **Hoshino, K., Inouye, H., Unokuchi, T., Ito, M., Tamaoki, N., and Tsuji, K.**, HLA and Diseases in Japanese, in *HLA and Disease*, INSERM, Paris, 1976, 249.
14. **Vermylen, C., Goethals, Th., and Van de Putte, I.**, Healthy carrier state and Australia antigen liver disease, *Lancet*, 1, 1119, 1972.
15. **Gyodi, E., Penke, S., Novak, E., and Hollan, S.R.**, HL-A specificities in individuals with persistence of hepatitis-associated antigenaemia, *Haematologia*, 7, 199, 1973.
16. **Jeannet, M. and Farquet, J.J.**, HL-A antigens in asymptomatic chronic HB Ag carriers, *Lancet*, 2, 1383, 1974.
17. **Seignalet, J., Robinet-Levy, M. and Lemaire, J.M.**, Les antigènes HL-A chez les portems sains de l'antigène Australia, *Nouv. Rev. Fr. Hematol.*, 14, 89, 1974.
18. **Boettcher, B.**, cited in **Ryder, L.P. and Svejgaard, A.**, Associations Between HLA and Disease, report from the HLA and Disease Registry of Copenhagen, 1976.
19. **Galbraith, R.M., Eddleston, A.L., Smith, M.G., Williams, R., McSween, R.N., Watkinson, G., Dick, H., Kennedy, L.A., and Batchelor, J.R.**, Histocompatibility antigens in active chronic hepatitis and primary biliary cirrhosis, *Br. Med. J.*, 3, 604, 1974.
20. **Bailey, R.J., Krasner, N., Eddleston, A.L., Williams, R., Tee, D.E., Doniach, D., Kennedy, L.A., and Batchelor, J.R.**, Histocompatibility antigens, autoantibodies, and immunoglobulins in alcoholic liver disease, *Br. Med. J.*, 2, 727, 1976.
21. **Mayr, W.R., Penner, E., and Sayfried, H.**, HLA- antigens in Gilbert's syndrome, in *HLA and Disease*, INSERM, Paris, 1976, 162.
22. **Bomford, A., Eddleston, A.L.W.F., Williams, R., Kennedy, L., and Batchelor, J.R.**, HLA-A3 and idiopathic haemochromatosis, in *HLA and Disease*, INSERM, Paris, 1976, 153.
23. **Fauchet, R., Simon, M., Bourel, M., Genetet, B., Genetet, N., and Alexandre, J.L.**, Idiopathic haemochromatosis and HLA antigens, in *HLA and Disease*, INSERM, Paris, 1976, 157.
24. **Simon, M., Pawlotsky, Y., Bourel, M., Fauchet, R., and Genetet, B.**, Hemochromatose idiopathique, maladie associée à l'antigène tissulaire HL-3?, *Nouv. Presse Med.*, 4, 1432, 1975.
25. **Shewan, W.G., Mouat, S.A., and Allan, T.M.**, HL-A antigens in haemochromatosis, *Br. Med. J.*, 1, 281, 1976.
26. **Gastard, J., Gosselin, M., Alexandre, J.L., Messner, M., Fauchet, R., and Genetet, B.**, HLA-A and B antigens in chronic primitive pancreatitis, in *HLA and Disease*, INSERM, Paris, 1976, 159.
27. **Polymenidis, Z., Ludwig, H., and Gotz, M.**, Cystic fibrosis and HL-A antigens, *Lancet*, 2, 1452, 1973.
28. **Kaiser, G.I., Laszlo, A., and Gyurkovits, K.**, HLA antigens in cystic fibrosis: An association of B18 with the disease, in *HLA and Disease*, INSERM, Paris, 1976, 252.
29. **Falchuk, Z.M., Rogentine, G.N., and Strober, W.**, Predominance of histocompatibility antigen HL-A8 in patients with gluten-sensitive enteropathy, *J. Clin. Invest.*, 51, 1602, 1972.
30. **Stokes, P.L., Holmes, G.K.T., Asquith, P., Mackintosh, P., and Cooke, W.T.**, Histocompatibility antigens associated with adult coeliac disease, *Lancet*, 2, 162, 1972.
31. **Evans, D.A.**, Coeliac disease and HL-A8, *Lancet*, 2, 1096, 1973.
32. **Keuning, J.J., Pena, A.S., van Hooff, J.P., van Leeuven, A., and van Rood, J.J.**, HLA-Dw3 associated with coeliac disease, *Lancet*, 1, 506, 1976.
33. **Solheim, B.G., Ek, J., Thune, P.O., Baklein, K., Bratlie, A., Rankin, B., Thoresen, A.B., and Thorsby, E.**, HLA antigens in dermatitis herpetiformis and coeliac disease, *Tissue antigens*, 7, 57, 1976.
34. **Mann, D.L., Nelson, D.L., Katz, S.I., Abelson, L.D., and Strober, W.**, Specific B-cell antigens associated with gluten-sensitive enteropathy and dermatitis herpetiformis, *Lancet*, 1, 110, 1976.
35. **Mann, D.L., Abelson, L., Henkart, P., Harris, S. D., and Amos, D.B.**, Specific human B lymphocyte alloantigens linked to HL-A, *Proc. Natl. Acad. Sci. U.S.A.*, 72, 5103, 1975.
36. **Nelson, D.L., Falchuk, Z.M., Kasarda, D., and Strober, W.**, Gluten-sensitive enteropathy: correlation of organ culture behavior with HL-A status, *Clin. Res.*, 23, 254, 1975.
37. **Albert, E.D., Harms, K., Wank, R., Steinbauer-Rosenthal, I., Scholz, S.**, Segregation analysis of HL-A antigens and haplotypes in 50 families of patients with coeliac disease, *Transplant. Proc.*, 5, 1785, 1973.

38. **McNeish, A.S., Nelson, R., and Mackintosh, P.**, HL-A1, and 8 in childhood coeliac disease, *Lancet,* 1, 668, 1973.
39. **Ludwig, H., Granditsch, G., and Polymenidis, Z.**, HL-A8 and haplotype HL-A1-8 in coeliac disease, *J. Immunogenet.,* 1, 91, 1974.
40. **Dausset, J. and Hors, J.**, Some contributions of the HL-A complex to the genetics of human diseases, *Transplant. Rev.,* 22, 44, 1975.
41. **van Rood, J.J., van Hooff, J.P., and Keuning, J.J.**, Disease predisposition, immune responsiveness, and the fine structure of the HL-A supergene, *Transplant. Rev.,* 22, 75, 1975.
42. **Seah, P.P., Fry, L., Kearney, J.W., Campbell, E., Mowbray, J.F., Stewart, J.S., and Hoffbrand, A.V.**, A comparison of histocompatibility antigens in dermatitis herpetiformis and adult coeliac disease, *Br. J. Dermatol.,* 94, 131, 1976.
43. **Murtagh, T.J., Reen, D.J., and Greally, J.**, HLA-A1 and B8 in coeliac disease, in *HLA and Disease,* INSERM, Paris, 1976, 166.
44. **Mougenot, J.F., Polonovski, C., Sasportes, M.L., Schmid, M., and Hors, J.**, HLA and digestive intolerance to gluten and cow's milk proteins, in *HLA and Disease,* INSERM, Paris, 1976, 165.
45. **Koivisto, V.A., Kuitunen, P., Tiilikainen, A., and Akerblom, H.K.**, HLA antigens, especially B8 and Bw15, in patients with juvenile diabetes mellitus, coeliac disease, and both of these diseases, in *HLA and Disease,* INSERM, Paris, 1976, 128.
46. **Mallas, E., Mackintosh, P., Asquith, P., and Cooke, W.T.**, Transplantation antigens in inflammatory bowel disease. Their clinical significance and their association with arthropathy with special reference to HLA-B27 (W27), in*HLA and Disease,* INSERM, Paris, 1976, 161.
47. **Vachon, A., Gebuhrer, L., and Betuel, H.**, HLA antigens in ulcerative colitis and Crohn's disease, in *HLA and Disease,* INSERM, Paris, 1976, 170.
48. **Van den Berg-Loonen, E.M., Dekker-Saeys, B.J., Meuwissen, S.G.M., Nijenhuis, L.E., and Engelfriet, C.P.**, Histocompatibility antigens and other genetic markers in ankylosing spondylitis and inflammatory bowel diseases, *J. Immunogenet.,* 4, 167, 1977.
49. **Bergman, L., Lindblom, J.B., Safwenberg, J., and Krause, U.**, HL-A frequencies in Crohn's disease and ulcerative colitis, *Tissue Antigens,* 7, 145, 1976.
50. **Tsuchiya, M., Yoshida, T., Mizuno, Y., Kurita, K., Hibi, T., and Tsuji, K.**, HLA antigens and ulcerative colitis in Japan, in *HLA and Disease,* INSERM, Paris, 1976, 169.
51. **Minev, M. and Tzekov, G.**, HLA and acute appendicitis, in *HLA and Disease,* INSERM, Paris, 1976, 1964.
52. **Darke, C., Orellana, W., and Dodge, J.A.**, HLA types and ABO blood groups in infants with pyloric stenosis and their mothers, *Tissue Antigens,* 7, 189, 1976.
53. **Horton, M.A. and Oliver, R.T.**, HL-A and pernicious anemia, *N. Engl. J. Med.,* 294, 396, 1976.
54. **Mawhinney, H., Lawton, J.W., White, A.G., and Irvine, W.J.**, HL-A3 and HL-A7 in pernicious anemia and autoimmune atrophic gastritis, *Clin. Exp. Immunol.,* 22, 47, 1975.
55. **Zittoun, R., Zittoun, J., Seignalet, J., and Dausset, J.**, HL-A and pernicious anemia, *N. Engl. J. Med.,* 293, 1324, 1975.
56. **Whittingham, S., Youngchaiyud, U., Mackay, I.R., Buckley, J.D., and Morris, P.J.**, Thyrogastric autoimmune disease, *Clin. Exp. Immunol.,* 19, 289, 1975.
57. **Reunala, T., Salo, O.P., Tiilikainen, A., and Mattila, M.J.**, Histocompatibility antigens and dermatitis herpetiformis with special reference to jejunal abnormalities and acetylator phenotype, *Br. J. Dermatol.,* 94, 139, 1976.

VIII. INFECTIOUS DISEASES AND IMMUNIZATION RESPONSES

Although numerous diseases in other sections have some relationship to infections, such as the reactive arthridities that follow Shigella, Salmonella, and Yersinia infections, this particular section will deal with the immediate response to infectious agents as well as to immunizations. Some infections are also described in other sections as noted below.

A. Infectious Mononucleosis

HLA antigen frequencies in 40 patients with infectious mononucleosis showed no significant alteration from that of controls, 37% of whom had low EB-virus titers, or from that of EB-virus antibody-negative controls.[1] These data failed to confirm Morris's finding of an increased frequency of Bw35 (W5).[2]

B. Congenital Rubella and Rubella Immunization

Eighty-seven patients with congenital rubella, 53 adults of whom 23 were males and 34 children of whom 18 were males, showed an increased incidence of several HLA antigens.[3] A1 showed the greatest alteration with an increase to 40.8% in the entire group of 87 patients compared to 32% of controls. More significantly, A1 was increased to 63.3% of 30 adult females (corrected P<0.02). Although there was also an increase of B5 from 8.0 to 20.1% in 41 male subjects, the difference was not significant when corrected for the number of antigens tested. These authors also attempted to correlate the different frequencies of HLA antigens with seropositivity to rubella in different populations between the ages of 20 and 29 years. In the 29 geographically different adult populations that were sampled, the highest correlation coefficients found between rubella seropositivity and HLA antigens were for A1, 0.71; B8, 0.60; and A1 and B8, 0.68. A correlation did not exist for A3-B7, A2-B12, or A2-B7 combinations, suggesting that the A1-B8 association was due to these antigens alone and not to their Caucasoid background. The virtual absence of A1 and B8 antigens in the Japanese population coincided with a low incidence of congenital rubella in Japan and a frequency of only 34% seropositivity for Japanese. In general, the populations with a high frequency of A1 and/or B8 showed greater levels of seropositivity than did populations where these antigens were low or absent. These results indicated that A1 and B8 may be more important in the spread of rubella epidemics than climatic, geographic, or socioeconomic conditions. The fact that A1 and B8 as a phenotype were increased in adult females but not in adult males would argue against A1 and B8 being the products of strong immune response genes that contribute to better survival of these individuals, as proposed by Mackay.[4] It might be that the A1 or B8 antigen is a favorable receptor site for rubella virus. But because the A1 antigen has the more significant association and because there is linkage disequilibrium between A1 and B8, it is possible that the B8 increase is merely a passive one due to linkage disequilibrium.

Following immunization of 232 rubella-seronegative children with RA 27/3 rubella vaccine, the highest geometric-mean convalescent phase titer occurred in children with the HLA antigens B14 and Bw22, who had geometric-mean titers of 152 vs. 88 in those without B14 or Bw22 (P<0.01).[5] Although only 8% of the total group of patients had A28, 40% (four to ten) who had rubella titers greater than 512 had this same antigen. On the other hand, of the six children in whom serum conversion did not occur, A2 was present in four and B12 and Bw17 in three. In 37 sibling pairs, there was good concordance between their antibody responses to rubella vaccine, with only 4 pairs showing greater than a fourfold difference in titers. No correlation could be established between arthralgia, arthritis, or vaccine-related lymphadenopathy and any HLA antigen. In contrast to an earlier study by these same authors in which Bw16-positive individuals had a poor antibody response to live intranasal influenza A vaccination,[6] there was no such association in this study with rubella vaccine.

In general, there was no consistency in the antigen alterations (A1-B8) seen in congenital rubella and seropositivity, those associated with good antibody response to rubella vaccination (B14, Bw22, and A28), or those found with no antibody response (A2, B12, and Bw17.)

C. Influenza A Vaccination

In the course of clinical trials with a live attenuated intranasal influenza A vaccination, it was found that subjects with the HLA antigen Bw16 had a mean convalescent phase hemagglutination inhibiting antibody titer of 1:14 which was significantly lower

TABLE 15

Recrudescent Herpes Labialis

HLA Ag	Controls with Ag (%)	Patients with Ag (%)	Patients with Ag (n)	Patients total (n)	Relative risk	Ref.
A1	25.1	55.6	50	90	3.7	8

than the mean titer of 1:36 in subjects without Bw16 (P<0.001).[6] Bw16 occurred in 32% of the 25 subjects with a poor antibody response, but in only 5% of those with a good antibody response as measured in both serum and nasal secretions. The crucial point in this study was that the intranasal route of immunization required multiplication of the virus before antibody production, and it was this group in which the Bw16-positive individuals showed a poor antibody response. However, when inactivated influenza vaccine was given intramuscularly, individuals with Bw16 had antibody responses comparable to those with other antigens. Such a difference in responsiveness depending on the type and route of immunization would suggest that its basis may rest on Bw16 individuals being more resistant to infection with live virus. Another interpretation would be that infection might have occurred via the nasal route following live attenuated virus installation, but other immune mediators such as interferon or soluble cellular factors might have been produced so rapidly that humoral responses were actually diminished.

D. Recrudescent Herpes Labialis

The herpes viruses have an unusual propensity for establishing latent infections after the initial primary exposure. Reactivation of the virus may occur at intervals. In man, the cytomegalovirus, varicellar zoster, and especially herpes simplex virus Types 1 and 2 share this characteristic.

Russell's follow-up study of his original observation that there was an increase in A1 in those individuals with a susceptibility to recurrent infection with herpes virus Type 1 included 260 patients in whom the frequency of A1 was increased to 50.1% compared to 23% of controls (Table 15).[7,8] This was the first demonstration of an association of a known viral disease with an HLA antigen.

Despite attempts with several techniques to detect a deficient immune response in patients suffering from recurrent herpes infection, none was found. On the contrary, all such patients have active humoral and cellular immunity to the viral antigens. However, the mechanisms of viral latency as opposed to initial viral infection are still unknown, and it may be that an active humoral or cell-mediated immune response may actually encourage viral latency.

E. Poliomyelitis (see Neurology and Psychiatry [Chapter 7, Section 10])

F. HB$_s$Ag Elimination and Antibody Production (see Gastroenterology [Chapter 7, Section VII])

G. Cytomegalovirus Infection

Of 78 renal allograft recipients who were studied prospectively, 56 acquired cytomegalovirus (CMV) infections.[9] An examination of the HLA antigens of those with CMV showed an increase in Aw32 (16% compared to 6% of controls), but its significance was lost when corrected for the number of antigens tested. Similarly, there was no significant HLA antigen deviation in the subgroup with CMV viremia nor in those who had an equivalent exposure but did not become infected. Furthermore, CMV infections occurred in 8 of 12 HLA identical siblings as well as an identical twin.[9] In ten of these HLA-identical donor-recipient combinations where CMV developed, there were five concordant and five discordant pairs.

H. Haemophilus Influenza — Type b

Sixty-five patients with *Haemophilus influenza* type b infections (epiglottitis or meningitis) showed the following significant differences in HLA antigen frequencies: in the A series, A11 occurred in 3% of the 30 patients with epiglottitis and 17% of the 35 patients with meningitis compared to 9.3% of controls; A28 in 20, 14, and 6.2% respectively; B5 in 13, 3, and 11%, respectively; Bw35 (W5) in 20, 29, and 11.4%, respectively; B14 in 13, 11, and 6.7%, respectively; B17 in 13, 11, and 3.2%, respectively; and B40 (W10) in 10, 23, and 15.3%, respectively.[10] The most outstanding differences were the increased frequency of B17 in both epiglottitis and meningitis patients ($P<0.02$), and of A28, predominantly in those with epiglottitis ($P<0.03$). A distinction occurred in the deviation of the antigens A11 and B40, which were higher in the meningitis patients, and in B5, which was higher in the epiglottitis patients. Other differences in the antigen profile relating to the production of antibody to the polysaccharide capsular antigen of *H. influenza* type b were uncertain, though patients with meningitis had distinctly lower and in fact, subnormal levels of antibody, whereas the epiglottitis patients had significantly higher levels than either controls or meningitis patients. On the one hand, whether the lack of any single antigen deviation or on the other, whether the emergence of a profile of antigen alterations similar to that seen with psoriasis will be confirmed, is yet to be seen.

A study of eight children with invasive *H. influenza* type b disease (seven with meningitis and one with epiglottitis; five Caucasians and three of Spanish origin) showed differences in the frequencies of both A- and B-series antigens in the Caucasian individuals and in the B series of antigens among the Spanish children.[11] Specifically, in the five Caucasian patients, A2 increased from 43.1% of 450 normal controls to 80% and B12 increased from 23.2% in controls to 40%. In the 3 Spanish patients, B12 increased from 21% of 100 normal controls to 66%, namely 2 out of 3 patients. These data suggest a possible influence of these antigens on the development of an invasive type of a disease following *H. influenza* b infection, but the sample size is very small.

I. Streptococcal Antigen Responses (including Rheumatic Fever and Heart Disease)

Lymphocytes from 217 random individuals when tested for their blastogenic response to Streptococcal antigens (streptokinase-streptodornase [SK/SD] with nuclease A) exhibited a high responder group in which B5 was significantly increased ($P<0.0001$).[12]

The response pattern to this purified Streptococcal antigen was found to be associated with HLA haplotypes in 25 families. In families with known recombinations between the B and D loci, the gene controlling this immune responsiveness appeared to be closely related with the B locus and to have occurred in linkage disequilibrium with B5.[13]

Another aspect of Streptococcal infections is the development of rheumatic fever and rheumatic heart disease, also discussed in Section VII.B. In 76 such patients, Falk found only a decrease in A3 and an increased frequency of shared antigens in their parents.[14] Although there was also an increase in B5 from 9.0 to 15.8%, this was not significant. A complete absence of A3 was reported in 48 such patients, but no comment was made about B5.[15] In 100 other patients, 50 Maoris and 50 Europeans, there were small increases in A3 and B8 in the former group and an increase in B17 and decrease in A28 in the latter group.[16] No difference was found for B5. Ward reported that among those with rheumatic heart disease, there was an increase in A29 and Aw30/31 in 58 patients without a history of rheumatic fever or chorea and no impressive differences in the 75 patients with such a history.[17] In Trinidad, an area endemic for Streptococcal infections, no correlation was found between rheumatic fever or nephritis and any HLA antigen.[18] Furthermore, 43 HLA-identical sibling pairs were discordant for the occurrence of Streptococcal disease. A significant decrease in B5 was detected only in mothers of patients.[18] Thus, to date, there is no pervasive correlation between the in vitro response to Streptococcal antigen, at least three clinical diseases related to the Streptococcus, and HLA antigens.

J. Leprosy

Leprosy affects only a small proportion of those who become exposed to *Mycobacterium leprae* because the great majority of exposed individuals develop effective immunity. In the low-resistance form of the disease called lepromatous leprosy, there appears to be a specific lack of cell-mediated immunity, whereas in the tuberculoid form, there is a normal resistance to the bacteria and the disease appears primarily as one of the nervous system.

Most of the HLA association studies in leprosy have involved unrelated individuals in different populations and have yielded equivocal results. Thorsby found an elevated frequency of Bw21 in Ethiopians,[19] Escobar-Gutierrez a decrease of both A2 and A3 in Mexicans,[20] Reis an increase of B5 and B27 in tuberculoid leprosy and Aw23 in lepromatous leprosy in Brazilians,[21] Kreisler an elevated frequency of B14 in Spaniards,[22] Smith an increased A10 in lepromatous and B5 in tuberculoid Filipinos,[23] and Dasgupta an increased B8 in lepromatous and a decreased A9 in nonlepromatous leprosy in Asian Indians.[24]

However, a recent study of 16 families with leprosy has strongly suggested that linkage exists with HLA.[25] Siblings with the same form of leprosy exhibited a significant excess of identical HLA haplotypes, a finding also occurring in families with only the tuberculoid form. Siblings with different forms of leprosy shared a haplotype less often than expected. The authors interpreted these data as showing that both susceptibility to and the form of leprosy were controlled by at least two HLA-linked genes. They further suggested that previous population studies had equivocal results because the HLA-linked genes had different linkage disequilibria with HLA-A and B loci in the various populations.[25]

K. Gonococcal Urethritis

In a small group of patients with various types of urethritis, A29 was found in 2 of 10 (20%) with gonococcal urethritis and in 6 of 23 (26%) with recurrent urethritis, compared to 6% of 19 patients who had neither gonococcal urethritis or recurrent urethritis, 6% in nonspecific urethritis, and 7% in 1085 healthy controls.[26] In recurrent urethritis, all six of the individuals with A29 also had B12, which is known to have linkage disequilibrium with A29. Whether the A29-B12 haplotype contained the susceptibility gene or whether one or the other of these antigens was found because of linkage disequilibrium is uncertain at the present time. Some diagnostic importance may be found in the fact that the B27 antigen was found in a distribution opposite to that of A29 and did not occur in a single patient with recurrent urethritis.

L. Gonococcal Arthritis (see Rheumatology [Chapter 7, Section XIV])

M. Periodontitis

In 47 patients with periodontitis, A2 was significantly decreased (21%), especially in the 16 females tested (12.5%). [27] But in the same paper, a preliminary report of another 72 patients failed to confirm this initial finding. Whether or not A2 is a protective gene product against periodontitis, as suggested by Asian Indians with a low A2 frequency and high periodontitis, is uncertain at present.

N. Immunization to rh (D)

In 62 responders and 31 nonresponders to rh (D), no significant alteration in HLA antigen frequency was noted.[28]

O. Tuberculin Hyporesponsiveness in BCG Treatment (see Allergy [Chapter 7, Section I])

P. Carcinoma of the Cervix (see Malignancy [Chapter 7, Section IX])

Q. Vaccinia Virus

Among 79 persons who received a primary vaccination, those with Cw3 had a significantly lower in vitro response to the virus as measured by lymphocyte transformation.[29]

REFERENCES

1. **Schiller, J. and Davey, F.R.**, Human leukocyte locus A (HL-A) antigens and infectious mononucleosis, *Am. J. Clin. Pathol.*, 62, 325, 1974.
2. **Morris, P.J. and Forbes, J.F.**, HL-A in follicular lymphoma, reticulum cell sarcoma, lymphosarcoma, and infectious mononucleosis, *Transplant. Proc.*, 3, 1315, 1971.
3. **Honeyman, M.C., Dorman, D.C., Menser, M.A., Forrest, J.M., Guinan, J.J., and Clarke, P.**, HL-A antigens in congenital rubella and the rule of antigens 1 and 8 in the epidemiology of natural rubella, *Tissue Antigens*, 5, 12, 1975.
4. **Mackay, I.R. and Morris, P.J.**, Association of autoimmune active chronic hepatitis with HL-A1, 8, *Lancet*, 2, 793, 1972.
5. **Spencer, M.J., Cherry, J.D., Terasaki, P.I., Powell, K.R., Sumaya, C.V., and Marcy, S.M.**, Clinical and antibody responses following immunization with RA 27/3 rubella vaccine analyzed by HL-A type, *Pediatr. Res.*, 10, 393, 1976.
6. **Spencer, M.J., Cherry, J.D., and Terasaki, P.I.**, HL-A antigens and antibody response after influenza A vaccination, *N. Engl. J. Med.* 294, 13, 1976.
7. **Russell, A.S.**, HLA and herpes simplex virus, type 1, in *HLA and Disease*, INSERM, Paris, 1976, 116.
8. **Russell, A.S. and Schlaut, J.**, HL-A transplantation antigens in subjects susceptible to recrudescent herpes labialis, *Tissue Antigens*, 6, 257, 1975.
9. **Braun, W.E., Nankervis, G., Banowsky, L.H., Protiva, D., Biekert, E., and McHenry, M.C.**, A prospective study of cytomegalovirus infections in 78 renal allograft recipients, *Proc. Clin. Dialysis Transplant Forum*, 6, 8, 1976.
10. **Robbins, S. J., Schneerson, R., Argaman, M., and Handzel, Z. T.**, Haemophilus influenzae type b: Disease and immunity in humans, *Ann. Intern. Med.*, 78, 259, 1973.
11. **Tejani, A., Fotino, M., Khan, R., Dobias, B., and Nangia, B.**, HLA and Hemophilus influenzae disease, in *HLA and Disease*, INSERM, Paris, 1976, 264.
12. **Greenberg, L.J., Gray, E.D., and Yunis, E.J.**, Association of HL-A5 and immune responsiveness in vitro to Streptococcal antigens, *J. Exp. Med.*, 141, 935, 1975.
13. **Greenberg, L.J., Chopyk, R.L., Ferrara, G.B., and Yunis, E.J.**, Genetic control of immune responsiveness to a Streptococcal antigen, in *HLA and Disease*, INSERM, Paris, 1976, 300.
14. **Falk, J.A., Fleischman, J.L., Zabriskie, J.B., and Falk, R.E.**, A study of HL-A antigen phenotype in rheumatic fever and rheumatic heart disease patients, *Tissue Antigens*, 3, 173, 1973.
15. **Gorodezky, C.**, HLA and rheumatic heart disease, in *HLA and Disease*, INSERM, Paris, 1976, 34.
16. **Caughey, D.E., Douglas, R., Wilson, W., and Hassall, I.B.**, HL-A antigens in Europeans and Maoris with rheumatic fever and rheumatic heart disease, *J. Rheumatol.*, 2, 319, 1975.

17. **Ward, C., Gelsthorpe, K., Doughty, R.W., and Hardisty, C.A.**, HLA antigens and acquired valvular heart disease, *Tissue Antigens,* 7, 227, 1976.
18. **Read, S. E., Reid, H., Poon-King, T., Fischetti, V. A., Zabriskie, J.B., and Rapaport, F. T.**, HLA and predisposition to the non-suppurative sequelae of Group A Streptococcal infections, *Transplant. Proc.,* 9, 543, 1977.
19. **Thorsby, E., Godal, T., and Myrvang, B.**, HL-A antigens and susceptibility to diseases. II. Leprosy, *Tissue Antigens,* 3, 373, 1973.
20. **Escobar-Gutierrez, A., Gorodezky, C., and Salazar-Mallen, M.**, Distribution of some of the HL-A system lymphocyte antigens in Mexicans, *Vox Sang.,* 25, 151, 1973.
21. **Reis, A.P., Maia, F., Reis, V.F., Androde, I.M., and Campos, A.A.**, HL-A antigens in leprosy, *Lancet,* 2, 1384, 1974.
22. **Kreisler, M., Arnaiz, A., Perez, B., Cruz, E.F., and Bootello, A.**, HL-A antigens in leprosy, *Tissue Antigens,* 4, 197, 1974.
23. **Smith, G.S., Walford, R.L., Shepard, C.C., Payne, R.O., and Prochazka, G.J.**, Histocompatibility antigens in leprosy, *Vox Sang.,* 28, 42, 1975.
24. **Dasgupta, A., Mehra, N.K., Ghei, S.K., and Vaidya, M.C.**, Histocompatibility antigens (HL-A) in leprosy, *Tissue Antigens,* 5, 85, 1975.
25. **De Vries, R.R.P., Nijenhuis, L.E., Fat, R.F.M.L.A., and van Rood, J.J.**, HLA-linked genetic control of host response to mycobacterium leprae, *Lancet,* 2, 1328, 1976.
26. **Rachelefsky, G.S., Osborne, M., and Terasaki, P.I.**, Frequency of HLA specificities in patients with a presumptive diagnosis of gonococcal urethritis, in *HLA and Disease,* INSERM, Paris, 1976, 260.
27. **Terasaki, P.I., Kaslick, R.S., West, T.L., and Chasens, A.I.**, Low HL-A2 frequency and periodontitis, *Tissue Antigens,* 5, 286, 1975.
28. **Dausset, J. and Hors, J.**, Some contributions of the HL-A complex to the genetics of human diseases, *Transplant, Rev.,* 22, 44, 1975.
29. **deVries, R. P., Kreeftenberg, H. G., Loggen, H. G., and van Rood, J. J.**, In vitro immune responsiveness to vaccinia virus and HLA, *N. Engl. J. Med.,* 297, 692, 1977.

IX. MALIGNANCY*

The most relevant antigens in malignancies are basically of three types: the first type is the histocompatibility antigens; the second is the possibly related Ia and Ir antigens, which may indicate more directly an individual's genetic susceptibility or resistance to the development of malignancy; and the third category of antigens are those tumor-specific antigens which represent the actual determinants on the cell surface membrane probably caused by whatever agent(s) produces the malignancy and are therefore not likely to be detected before the individual has been afflicted with the disease. Consequently, the antigens in the first two groups would be most important in ascertaining susceptibility, resistance, and risk for malignancy, whereas those in the third group would indicate the presence, absence, or perhaps degree of activity of the disease. These antigens might eventually constitute a profile to be used together when evaluating the occurrence and prognosis of various malignancy types. However, only the first two types of antigens will be discussed here.

A. Aplastic Anemia**

In a study of 100 patients with aplastic anemia (62 males and 38 females), 86 of the idiopathic variety, 5 following chloromycetin, and 9 in association with liver disease, 62% were found to have B12 compared to 44% of 591 controls ($P < 0.03$).[1] In a study of the families of 38 of these patients, there was a greater than expected frequency of A-series antigen homozygosities (13 observed compared to 8 expected, $P < 0.01$), a

* Additional data on HLA and malignancy may be found in a recent publication, *HLA and Malignancy,* Murphy, G. P., Ed., Alan R. Liss, New York, 1977.

** This disease is included under Malignancy because HLA reports on it have generally been included with hematologic malignancies.

finding that also occurred in patients with acute lymphocytic leukemia.[1] However, analysis of 200 such patients failed to show any significant association with any HLA antigen or haplotype, but the question of increased homozygosities of A-series antigens was not addressed.[2]

B. Carcinoma

1. Carcinoma of the Bladder

Studies by Takasugi showed a borderline decrease of 13% in Bw35 in 139 patients compared to 22% in controls ($P < 0.05$).[3]

2. Carcinoma of the Breast

Several studies have been conducted in search of an association between an HLA antigen and carcinoma of the breast. The initial study in 1972 by Patel suggested a significant association with the antigen B7.[4] In 52 patients with carcinoma of the breast, 38.5% had B7 compared to 19.5% of 123 controls. However, these authors could find no relationship of B7 to the age of disease onset, the metastasis-free interval, or the frequency of development of metastases. Another early study by Cordon in 97 patients failed to find significant differences in any HLA antigens when compared with 107 healthy controls.[5] The antigen B7 was actually found in more controls (30%) than patients (20%). A study of a broad variety of cancers by Takasugi[3] also showed no significant alteration in HLA antigens in 384 patients with breast carcinoma as did another study of 200 patients and matched controls.[6] Other conflicting data also were provided by Martz, who reported that in 48 Boston-area patients (where the initial positive study of Patel was performed) with carcinoma of the breast, no significant increase in B7 frequency could be found.[7] This study was conducted with a total of 122 controls, predominantly females, matched even for the country of origin of the patient's ancestors and race. This study showed that B7 occurred in only 23% of the 40 patients with breast cancer compared to 19% of matched controls. Only an insignificant increase in the frequency of B14 from 6% of controls to 15% of patients and a decrease in B12 from 25% in controls to 13% in patients were noted. All pairs of antigens and all possible haplotypes failed to reveal any differences. It is interesting to note that even at the date of publication of this report, one of the reasons selected by these authors for the difference in the initial positive studies of Patel and their own negative findings was a difference based on the adequacy of the typing sera used to type an antigen supposedly as well defined as B7. In a very recent study of 150 patients with breast carcinoma in which the patient population was divided into various clinical subgroups based on whether the patients were pre- or postmenopausal and whether they had or had not been pregnant, it was found that in those women with A28 or Bw35 who were nulliparous and postmenopausal, there was an increased frequency of carcinoma of the breast.[8]

It should be appreciated that an increased risk for a disease can be found simply on clinical analyses of different groups of patients with breast carcinoma. For example, Anderson has found that there was a 40-fold risk of breast carcinoma for women between the ages of 20 and 29 whose mothers had had cancer and who were sisters of patients with premenopausal and bilateral cancer.[9] Such families offer promising material for even more revealing analyses of HLA and Ia antigens.

3. Carcinoma of the Bronchus

A positive association between the presence of antigens Aw19 and/or B5 and a 1-year disease-free survival in a prospectively studied group of 70 Caucasian patients with squamous and adenocarcinoma of the lung was further evaluated with 2-year survival data.[10] Of 69 patients who were available for follow-up, 18 were disease-free and 51 were either dead of their disease or had metastases. Of these 69 patients, 9 had Aw19 and 5 were free of disease at 2 years. Of the 12 who possessed B5, 7 were disease-free at 2 years. Of the 48 patients who had neither of these antigens, only 6 were disease-free. Therefore, 57% (12 of 21 patients) having either of these two antigens were alive or disease-free at 2 years compared to 13% (6 of 48) of patients who lacked these antigens ($P < 0.001$). Of the six disease-free patients who did not have either Aw19 or B5, four had the B5 cross-reacting antigen Bw35, thereby leaving only two patients who lacked some member of this group of antigens. It is extremely interesting that these antigens, Aw19 and B5, are the same two antigens which have been found to be associated with a much *poorer* prognosis in Hodgkin's disease. This would suggest that the etiologic agent responsible for these neoplasms operates through different genetic resistance or susceptibility mechanisms.

4. Carcinoma of the Cervix

This disease, which is believed to have a significant relationship to herpes genitalis type 2 infection, provides a means of assaying both the susceptibility to a virus infection as well as the susceptibility to develop a malignant disease. In a study of 42 patients with cervical carcinoma, 36 female patients with recurrent herpes genitalis, and 450 controls, the only difference noted was an increase in the antigen B15 in the cervical carcinoma patients when compared to the controls (12 of 42, or 29% compared to 43 of 450, or 10% respectively; corrected $P < 0.02$).[11] The patients with recurrent herpes genitalis, who are believed to have a higher risk of developing cervical carcinoma, showed no significant difference in their HLA antigen frequencies when compared to the controls or to the cervical carcinoma patients. In another study of 121 females with carcinoma of the cervix compared to 212 healthy females, it was found that B12 was increased in those with localized but especially those with invasive carcinoma.[12]

5. Carcinoma of the Colon

In a study of 121 patients, 14% had a borderline decrease in Bw35 compared to 22% of controls ($P < 0.05$).[3]

6. Carcinoma of the Endometrium

A study of 68 patients showed no significantly altered HLA antigen frequency.[3]

7. Carcinoma of the Esophagus

Two preliminary studies from Iran showed only a suggestive increase of B40 in 58 patients from different ethnic groups in one study[13] and in the other study of 43 patients, a decrease in two antigens, HLA-A9 from 25 to 7% and B5 from 34 to 15%.[14]

8. Carcinoma of the Kidney (see Renal Diseases [Chapter 7, Section XIII])

Although there has been no association between any HLA antigen and renal cell carcinoma in a random population, [3,15] a study of familial cases has shown a striking association between familial renal cell carcinoma and B17.[15]

9. Carcinoma of the Larynx

One hundred and fifty patients with carcinoma of the larynx were found to have no significant deviation as a group of any HLA antigen. Certain subgroups did appear to show some differences based on whether the carcinoma had been cured for at least 3 years.[16] This report lacked specific data.

10. Carcinoma of the Nasopharynx

In 31 patients from Singapore with nasopharyngeal carcinoma, there was found to be an increased frequency of A2 (61.3% compared to 45.2% of controls).[17] A high number of blank alleles were found in the B-series antigens which could be accounted for by a new second-series specificity that was designated SIN-2 for Singapore 2. It occurred in nine of ten proved cases of nasopharyngeal carcinoma. This antigen has also been reported by Payne as Hs.[18] Thus, an antigen which seems to be rather restricted to Orientals, particularly the Chinese, may very well have quite a high frequency in those individuals with nasopharyngeal carcinoma, a type much more frequent in the Oriental population. A new D-series antigen designated Sin-2a has very recently been found in virutally 100% of the patients tested.[19]

11. Carcinoma of the Ovary

A single study revealed that 69 patients had no significant deviation in HLA antigen frequency.[3]

12. Carcinoma of the Prostate

In a study of 214 patients, no significant differences in HLA antigen frequencies were found.[3]

13. Carcinoma of the Rectum

In a study of 55 patients, there was an increase in A9 from 21% in controls to 36% in patients ($P < 0.05$).[3]

14. Carcinoma of the Stomach

Early studies by Takasugi showed no significant alteration in HLA antigen frequency.[3] In a study of 87 gastric carcinoma patients from Japan (where the disease is quite prevalent), there was an increased frequency of A9.[20,21] When sera were tested from patients with gastric carcinoma, 60% (9 of 15) showed B-cell activity but no T-cell antibody. Whether this antibody was related to the primary disease or to transfusions given in the course of surgical treatment of the disease is unknown. Furthermore, the relationship of B-cell antibody to carcinoembryonic antigen which may occur in these patients and which we have noted to show a high association with B-cell antibody production has not been explored.[22]

C. Leukemia

1. Acute Lymphocytic Leukemia (ALL)

Just as in Hodgkin's disease, one of the most remarkable findings is the general lack of association of ALL with any HLA antigen.[23] However, there have been suggestions of a low degree of increased risk with several HLA antigens, namely, A2, B8, and B12. Beginning with the early work of Kourilsky in 1967,[24] many studies primarily found an increase in A2. The strongest report concerning A2 came from Rogentine in 1972 when he reported a 72% frequency of A2 in 50 patients with acute lymphocytic leukemia compared to 41% in 200 controls.[25] At that time, numerous other control groups had shown substantially higher percentages of A2: for instance, 53% of 234 controls reported by Kourilsky,[24] 45% of 308 controls reported by Dausset,[26] 50% of 1028 controls reported by Albert,[27] and 49% of 180 controls reported by Pegrum.[28] Thus, the very high A2 frequency in Rogentine's patient group together with the lowest reported frequency for A2 in any control group, 41%, combined to provide a significant relationship between ALL and A2.

Dausset emphasized that the increased frequency of A2 in ALL is a consequence of studies done retrospectively and thereby biased by patient survival and by the antigens prevalent in the long-lived, but not necessarily all, patients with ALL.[26] When Dausset examined 100 of his original cases of ALL, there was a slight but significant increase in A2 in the survivors ($P < 0.05$), though there was no difference in the 84 patients who were under 16 years old when the diagnosis was made. He noted that in Rogentine's study in 1973, of his 32 cases studied retrospectively with the diagnosis established between 1962 and 1968, 84% had A2 in sharp contrast to the 53 cases diagnosed between 1968 and 1971 in which the A2 frequency decreased to only 53%, a level not significantly different from the 44% in normal controls.[29]

However, when Rogentine grouped A2 with strong and weak cross-reactors for this antigen, namely, A28 and A9, he found that 91% of the early diagnosed cases possessed one of these three antigens, 94% of those who survived more than 1500 days had one of these antigens, and 94% of those with no extremes in their white counts and apparently a less serious form of the disease possessed one of these three antigens.[29] All of these associations were significant and supported the contention that these antigens were associated with a better prognosis. In Lawler's study in 1974, there was an increase in A9 but no significant increase in A2 or A28 in survivors.[30] On the other hand, there was a much shorter median survival in the patients with A10 (340 weeks compared to 100 weeks). Whether this type of antigen association with resistance or more appropriately, a relative resistance, is analogous to that based on the Rgv-1 gene (resistance to Gross virus) described by Lilly in mice is still highly speculative.

A recent study in 126 patients with acute lymphocytic leukemia showed that A2 was increased to 57% compared to 44% in 591 controls.[1] In addition, B5 was also increased to 23% compared to 13% of controls. There was also an increase in homozygosity for the A-locus antigen in 24 informative families in which there were 10 observed homozygosities in the A series when only 6 were to be expected.

In a very large *prospective* report of 309 North American Caucasians, Johnson found no significant difference in antigen frequency at either the A or B locus.[31] However, in the 251 patients who were also typed for C-locus specificities, it was found that Cw3 was lower in ALL than in the control group (9 and 27%, respectively; corrected $P < 0.01$). Thus, these authors suggested that the association, if any does exist between an HLA antigen and ALL, is with a C-series antigen rather than one in the A or B locus and that the lower frequency of Cw3 indicates resistance to ALL.

Perhaps the most provocative study was that by Terasaki, who found that in 60 leukemic patients, which included 19 AMLs, 14 ALLs, 8 CMLs, 7 CLLs, and 12 unclassified leukemias, there was a complete absence of one of the new B-cell antigen groups, Group 2, in all of the leukemics, compared to a frequency of approximately 23% in controls.[32] Furthermore, this antigen was lacking on 37 cultured lymphoblastoid lines. Terasaki's groups B1 and B2 are believed to be gene products of the first B-lymphocyte locus, whereas the other three B antigens, B3, B4, and B5, are thought to be the gene products of the second locus, both of which are believed to be in the HLA chromosomal region. The group 1 antigen, though not totally absent in leukemia patients, was found to have a lower frequency of 18% compared to 48% of controls. The absent or low frequency of these two known antigens of the first B-lymphocyte locus, Group 2 and Group 1, respectively, was used as evidence of an increased frequency of some as yet undetected first B-locus antigen. These authors previously reported that approximately 70% of AML, ALL, and CML and almost all CLL express Ia-like B-lymphocyte alloantigens. Since the authors examined only those leukemic cells that expressed B-cell antigens, that is, they did not include T-cell ALL, they thought it was possible that the T-cell ALLs could express entirely different frequencies of the B antigens (known and unknown) from those reported in this study for the non-B, non-T leukemias. These data suggested that an antigen either within the first B-lymphocyte locus or linked to it might constitute the gene product of a resistance gene to leukemia, though other possibilities were considered. Since almost all of the B-cell lines that were tested were infected with Epstein-Barr virus, it was possible that those individuals who had the B Group 2 were resistant not only to leukemia but also to B-cell transformation with Epstein-Barr virus. Unfortunately, the validity of the Terasaki B cell Group 2 is now in serious doubt, and the whole question of resistance to leukemia in those having this antigen will have to be reevaluated.

Thus, one might consider patients exposed to leukemic agents, viruses, or other agents in three groups:

1. Those who are totally resistant to getting the disease, such as those with B-cell Group 2 of Terasaki, if such a group is substantiated
2. Those that are relatively resistant (possibly heterozygotes) to getting the disease in its worst form and therefore have the best prognosis, such as those with A2, A28, A9, and Cw3
3. Those who have the greatest likelihood not only of getting the disease but also of showing the most aggressive form of the disease because of their genetic susceptibility

Surprisingly, no antigen has been found so far to indicate increased susceptibility — only the resistance-related antigens noted above have been found. However, the suggestion that homozygosity of a child due to shared parental antigens may be associated with an increased leukemia risk corresponds with the concept that heterozygosity is a protective mechanism. (See Possible Mechanisms of HLA and Disease Associations [Chapter 3]).

2. Acute Myelogenous Leukemia

Benbunan found no significant difference in HLA antigen frequencies in 79 patients with acute myelogenous leukemia and 79 with chronic myelogenous leukemia when compared to a control population of 591 healthy people.[33] Oliver's study of 103 patients with AML and a comparable control population showed no significant antigen differences overall.[34] Yet, when survival for more than 2 years in 22 patients was compared to that of 62 patients who died in the first year, it was found that there was a significant increase of B12 in the long-term survivors (50% compared to 21% for short survivors and 27% for controls). Examining this group of long-term survivors for possible interaction between the A- and B-locus gene products, they found that there was a significantly better survival of patients who were positive for A1-B8, A2-B12, or A3-B7, which suggested some phenotypic interaction.

Studies of granulocytic antigens including six allelic specificities detected on normal granulocytes showed that granulocytic antigens occurred less frequently on granulocytes of patients with acute myelogenous leukemia than on normal granulocytes.[35] Interestingly, two sera with no known specificity reacted more frequently against leukemic cells, a finding which suggested possibly a new antigen group with greater significance or greater specificity for acute myelogenous leukemia.

D. Lymphoma

1. Burkitt's Lymphoma

In a combined analysis of three independent studies of 106 patients with Burkitt's lymphoma, the only two differences that approach significance at the 5% level were an increase in A11 and B14.[36] This analysis, which very carefully evaluated minor differences in antigens occurring in each of the component studies, provided a valuable statistical approach toward combining studies and evaluating the heterogeneity among them.

2. Hodgkin's Disease

Numerous studies of HLA antigens in Hodgkin's disease have been done through the years and, in fact, were among the earliest to indicate a possible association with an HLA antigen.[37-46] The antigens that were increased in the majority of these studies were multiple ones including those related to the B5 cross-reacting group (B5, B18, and Bw35, earlier called 4c), A1, B8, as well as A9 and A10.[46]

Some interesting subgroups of patients were found in these studies. For example, Falk noted that there was an increased frequency of B8 in those whose disease had a duration greater than 5 years as well as in the lymphocytic type, whereas there was an increase in A1 in the mixed cell type.[41] Forbes, in addition to finding an increased frequency of Bw35, then known as W5, also noted an increase in A11 in females with the nodular sclerosing type.[39,45] Increased frequencies of A1 and B8 were found again by Kissmeyer-Nielsen.[47] The same suggestion of an increase in A1 and B8 was also noted in the combined material of 523 patients utilized in the 5th International Workshop of Histocompatibility Testing.[48] In this study also, no convincing evidence was found for an increase of the 4c-related antigens, A5, Bw35, and B18.

In a study of a variety of hematological malignancies, Miller found that there were only suggestive increases in B5 and B13 in small groups of patients with lymphoma, multiple myeloma, chronic lymphocytic leukemia, Hodgkin's disease, polycythemia vera, and chronic myelogenous leukemia.[49] The increase in B13 was most pronounced in patients having lymphoma, multiple myeloma, and Hodgkin's disease, whereas the B5 increase was predominantly in myeloma, polycythemia vera, and less so in lymphoma.

Since most of these studies were primarily retrospective, the survival of patients might influence the antigen frequency. Consequently, a prospective study done by Kissmeyer-Nielsen was of potentially great value.[50] This prospective study of 80 female and 121 male patients showed that there were virtually no changes in antigen frequency of the HLA antigens compared to normals. There was only a slight increase in the A1 frequency to 44% compared to 34% of controls, and in B8, 26% compared to 23% of controls, yielding relative risks of 1.54 and 1.19, respectively.[50] This result was in partial agreement with an evaluation of combined series.[46] Twenty-six patients who died in the course of this study had a significant increase in the number of phenotypes with only one A-series antigen. As a consequence of this, there was a decreased frequency in antigens A9, A10, A11, and Aw19, so that as a group, these antigens occurred in only 15% of the deceased patients compared to 55% of the survivors and 50% in the controls.[50] B8 was found to be significantly increased in those patients whose onset of disease was after the age of 30 years. According to the histologic description of the lesions, A1 and B8 had the lowest frequency in patients with nodular sclerosis and the highest frequency in patients with mixed cellular lesions, but this was significant only for the B8 antigen. Furthermore, the A1 and B8 were more frequent among males than females whether or not the onset was early or late. These authors concluded that there was

1. An increase in A1 in patients with Hodgkin's disease though the relative risk was only slightly increased (1.53)
2. An association between A1 as well as B8 with sex, age of onset, and the histology of the lesions
3. An unusual absence, or homozygosity, of A-series antigens so that a group of antigens occurred much less frequently than in controls

Another prospective study of 65 Swedish cases of Hodgkin's disease also failed to show any striking HLA antigen frequency changes though there was a slightly increased frequency of B8 in patients over 40 years of age, a more marked increase in B12 in those under 40 years of age who had favorable histopathology, and an increase

TABLE 16

Malignant Melanoma

HLA Ag	Controls with Ag (%)	Patients with Ag (%)	Patients with Ag (n)	Patients total (n)	Relative risk	Ref.
B5	15.0	13.0	7	54	0.8	53
	11.0	14.0	7	50	1.3	3
	22.0	0.0	0	29	0.06	55
	13.7	17.1	6	33	1.4	54
	11.9	14.1	30	212	1.2	56
	9.5	6.1	2	33	0.6	57
B27	8.0	7.0	4	54	1.0	53
	8.0	12.0	6	50	1.6	3
	6.1	3.3	1	33	0.5	54
	8.4	7.1	15	212	0.8	56
	8.5	24.2	8	33	3.4	57
	11.0	24.0	13	54	2.6	58

in A28 in those with advanced disease or a poor response to PPD and pokeweed mitogen.[51]

Thus, the dilemma of HLA antigen associations with Hodgkin's disease remains unresolved. Numerous early studies suggested an association with the 4c antigen (B5 cross-reacting group), but later prospective studies showed virtually normal antigen frequencies with only a tendency to an A1-B8 increase that, nonetheless, appeared to have some clinical relevance.

A further complexity was introduced in a recent report to the American Society of Clinical Oncology by Osoba and Falk, who reported that in 79 patients followed prospectively for at least 30 months after the tissue diagnosis of Hodgkin's disease was made, antigens Aw19 or B5 singularly or together occurred more frequently than expected in those patients whose initial therapy failed to achieve complete remission, as well as in those who died within 2 years after diagnosis.[52] Of 24 patients who were positive for either Aw19, B5, or both, 9 died within 42 months, compared to only 7 of the remaining 55 patients. Only 13 patients possessing these antigens compared to 46 without them had a complete remission. The frequency of a complete remission in Aw19- and B5-*negative* patients was extremely high in those with Stage 2 or 3 disease (95%) and decreased sharply in patients with Stage 4 disease (57%, $P < 0.0001$). In addition, the overall remission rate of 57% in Aw19- and B5-*positive* patients was poor, regardless of the stage of the disease. These authors suggested that males who did not have Aw19 and/or B5 had the best chance of achieving a complete remission after initial therapy, regardless of the histology or the stage of the disease, while males with these antigens had the lowest chance of achieving a complete remission. The influence of these antigens was not as strong in a small number of females. [52]

E. Malignant Melanoma

Numerous studies have shown no consistent significant association between an HLA antigen and malignant melanoma (Table 16).[23] A study by Cordon of 54 patients who had an increase in B18[53] failed to substantiate an earlier report of 31 patients that demonstrated an increased frequency of A9. [54] Clark found a complete absence of B5 in 29 patients. [55] A study of 212 patients had several significant changes in HLA antigen frequencies: lower B7 overall and increased A1 and Aw19 in patients with 5-year survival rates of 0 to 19% and 20 to 79%, respectively.[56] An increase in B27[57] from 8.5% of 200 controls to 24.2% of 33 patients was confirmed in 54 Norwegian patients by Bergholtz, who reported that the frequency of B27 was increased to 24% compared to 11.0% in 788 normal controls.[58] Although the corrected P value was not significant, had these authors chosen to analyze just the B27 antigen suggested by Singal's study, the P value would have been < 0.005. These authors also investigated the D-series antigens by means of homozygous cell typing and found that there was no significant difference in the antigen frequency for Dw1,2,3,4,5, or 6, though there was an increased frequency of LD 108 (21% compared to 8.6% of controls; P < 0.05, uncorrected).[58]

F. Multiple Myeloma

One of the earliest studies by Bertrams described an incidence of 35% for an antigen then identified as the non-B5, non-Bw35 component of 4c, namely, B18.[44] This antigen occurred in only 9.1% of 255 controls. However, this component of 4c, B18, could not be correlated with any specific monoclonal protein type. A later study of 44 patients (32 Caucasians and 12 Negroes) with multiple myeloma again showed an increase in the frequency of the 4c complex of antigens in Caucasians with a frequency of 56% compared to 30% of 131 controls.[59] However, the differences here were made up primarily by B5 (16% of patients compared to 6% of controls), Bw35 (13% in patients compared to 4% of controls), and not at all by B18 (3% in patients compared to 6% of controls). No significance of 4c was found after correction. When the subtypes of myeloma were evaluated (IgA, IgG, or type unknown), the test groups, though quite small, showed suggestive associations in that 4 of 6 patients with the IgA type of myeloma were A9-positive, whereas only 3 of 17 patients with the IgG type had this antigen. [59] When subtypes were analyzed according to the type of light chain found in the serum and/or urine of the patients, an increase in Bw22 was found in the kappa subtype.

Another study showing the same grouping of increased antigen frequencies within the 4c complex came from England.[60] In 63 myeloma patients and 83 normal controls, B5 was found to be more frequent among the patients (15.9%) than among controls (3.6%), though the P value for this difference was not significant when corrected. In addition to the problem of serologically defining some of the components of the 4c complex, the other possible reason for an association with 4c antisera is the fact that these antisera may contain specificities not related to HLA but to other antigen systems, perhaps B-cell or Ia specificities.

Expanding on their earlier studies, Jeannet and Magnin investigated 170 patients with hematologic malignant disease, including 14 with multiple myeloma. In their small groups of patients, only a slight increase in B5 (36% compared to 22% in controls), a decrease in B18 (0% compared to 7% of controls), and a decrease in Bw35 (7% compared to 19% of controls) were found.[61] (See also discussion of paraproteinemia and monoclonal gammopathy under Complement Deficiencies and Immunodeficiency Diseases [Chapter 7, Section III]).

G. Retinoblastoma

A study of 19 patients who had a spontaneous regression of a retinoblastoma and who were therefore an interesting group to investigate for unique HLA antigen association, nevertheless showed no significant association with any HLA antigen of the A, B, or D series.[62] In families in which several members had had retinoblastoma, the occurrence of the tumor was shown not to segregate with HLA. Moreover, in several families in which one parent had shown spontaneous regression of the tumor, children inheriting either of the parental haplotypes had tumors that did not regress.

In a large series of 122 patients with retinoblastoma, Bertrams found that Bw35 was increased to 25.4% of the patients compared to 11% of 255 controls, whereas B12 was reduced to 9.8% in the patients compared with 25.1% of controls, thus yielding relative risks of 2.75 and 0.34, respectively.[63] These same two antigens were examined in a study by Gallie. [62] The B12 antigen frequency was not reduced and appeared in 22.2% of 18 patients with spontaneous regression, 18.2% of 55 patients without regression, and 20.6% of 855 controls. [62] Similarly, Bw35 occurred in 16.7% of the 18 patients with spontaneous regression, 14.6% of the 55 patients without regression, and 18.8% of controls. Consequently, there seems to be little evidence that the occurrence, spontaneous regression, or progression of retinoblastoma is associated with any HLA antigen, even in families.[62]

H. Trophoblastic Neoplasms

Because of the possibility that trophoblastic neoplasms in the female might be related to certain HLA antigens, to increased histocompatibility between the patients and the spouse that could lead to the lack of rejection of the trophoblastic tissue, or to the lack of immune response in spite of HLA incompatibility of the male spouse and of the fetus, these antigens were investigated in 111 Caucasian patients with trophoblastic disease.[64] No statistically significant differences in the phenotype frequencies of any of the HLA antigens were found when compared to 1259 healthy Caucasian controls. However, there was an increase in A11 in 39 patients who currently had the disease but not in 72 individuals who had recovered from the disease. There was also an increase in B18 among 18 patients who currently had invasive disease, such as choriocarcinoma or an invasive mole, but not in 44 individuals who had recovered from the disease. Such associations of A11 and B18 with the aggressive form of the disease are reminiscent of similar findings for antigens Aw19 and B5 in diseases such as Hodgkin's. Furthermore, no increase in histocompatibility was found in 45 patient-couples compared to 67 control-couples. Similar negative compatibility results were reported by Lawler in 53 patients in whom only a tendency for A1 and B8 increases was found.[65] In the context of the known higher frequency of choriocarcinoma in group A women with group O husbands, there was no influence of HLA compatibility.[66] Nevertheless, a good response to therapy seemed to be more likely in those with no compatibile husband-wife antigens than in those with one or more compatibilities, there being only one death in cases in the former group compared to 5 of 29 in the latter group.[65] Therefore, no firm data have been found for the higher prevalence of choriocarcinoma according to HLA type, though the prognostic aspect does appear to have merit.[67]

I. Changes in HLA Antigens in Patients with Certain Malignancies

Two interesting examples of HLA antigen loss on circulating cells in patients with malignancies have been reported.[68,69] In Seigler's case of lymphoma, HLA antigens detectable on skin fibroblasts and circulating cells during remission disappeared from the circulating cells in relapse.[68] Following treatment with ^{60}Co irradiation, Bertrams found that all HLA antigens except 4a and 4b were undetectable in a case of Hodgkin's disease.[69] The reciprocal relationship between HLA and tumor activity in Seigler's patient resembles the alternate presence of TL antigen and H-2D antigens in the process of modulation.[70]

REFERENCES

1. **Gluckman, E., Lemarchand, F., Nunez-Roldan, A., Hors, J., and Dausset, J.,** Possible excess of HLA-A homozygous among aplastic anemia and acute lymphoblastic leukemia (ALL), in *HLA and Disease,* INSERM, Paris, 1976, 226.

2. **Albert, E., Thomas, E.D., Nisperos, B., Storb, R., Camitta, B.M., and Parkman, R.,** HLA antigens and haplotypes in 200 patients with aplastic anemia, *Transplantation,* 22, 528, 1976.
3. **Takasugi, M., Terasaki, P.I., Henderson, B., Mickey, M.R., Menck, H., and Thompson, R.W.,** HL-A antigens in solid tumors, *Cancer Res.,* 33, 648, 1973.
4. **Patel, R., Habal, M.B., Wilson, R.E., Birtch, A.G., and Movre, F.D.,** Histocompatibility (HL-A) antigens and cancer of the breast, *Am. J. Surg.,* 124, 31, 1972.
5. **Cordon, A.L. and James, D.C.,** HL-A and carcinoma of the breast, *Lancet,* 2, 565, 1973.
6. **de Jong-Bakker, M., Cleton, F.J., D'Amaro, J., Keuning, J.J., and van Rood, J.J.,** HL-A antigens and breast cancer, *Eur. J. Cancer,* 10, 555, 1974.
7. **Martz, E. and Benacerraf, B.,** Lack of association between carcinoma of the breast and HL-A specificities, *Tissue Antigens,* 3, 30, 1973.
8. **Bouillenne, Cl. and Deneufbourg, J.M.,** Breast cancer, in *HLA and Disease,* INSERM, Paris, 1976, 219.
9. **Anderson, D.E.,** cited in, One disease concept seen veiling genetic risk in breast cancer, *Oncology News,* 2, 3, 1976.
10. **Rogentine, G.N., Jr., Dellon, A.L., and Chretien, P.B.,** Prolonged disease-free survival in bronchogenic carcinoma associated with HLA Aw19 and B5. A follow-up, in *HLA and Disease,* INSERM, Paris, 1976, 234.
11. **Dostal, V. and Mayr, W.R.,** HLA SD antigens in cervix cancer and in recurrent herpes genitalis patients, in *HLA and Disease,* INSERM, Paris, 1976, 223.
12. **Koenig, U.D. and Muller, N.,** Cervical carcinoma and HLA antigens, in *HLA and Disease,* INSERM, Paris, 1976, 228.
13. **Mohagheghpour, N., Dolatshahi, K., Hashemi, S., Modabber, F., and Takasugi, M.,** HLA-B40 in Iranian patients with oesophageal cancer, in *HLA and Disease,* INSERM, Paris, 1976, 307.
14. **Nikbin, B., Ala, F., Amiri, C.H., and Moghadam, J.A.,** HLA antigens in oesophagal carcinoma (OC), in *HLA and Disease,* INSERM, Paris, 1976, 230.
15. **Braun, W.E., Strimlan, C.V., Negron, A.G., Straffon, R.A., Zachary, A.A., Bartee, S.L., and Grecek, D.R.,** The association of W17 with familial renal cell carcinoma, *Tissue Antigens,* 6, 101, 1975.
16. **Deneufbourg, J.M. and Bouillenne, Cl.,** Larynx cancer, in *HLA and Disease,* INSERM, Paris, 1976, 222.
17. **Simons, M.J., Chan, S.H., Wee, G.B., and Shanmugaratnam, K.,** Probable identification of an HL-A second-locus antigen association with a high risk of nasopharyngeal carcinoma, *Lancet,* 1, 142, 1975.
18. **Payne, R., Radvany, R., and Grumet, C.,** A new second locus HL-A antigen in linkage disequilibrium with HL-A2 in Cantonese Chinese, *Tissue Antigens,* 5, 69, 1975.
19. **Simons, M.J., Chan, S.H., Ho, J.H.C., Chau, J.C.W., Day, N.E., and de The, G.B.,** A Singapore 2-associated LD antigen in Chinese patients with nasopharyngeal carcinoma, in *Histocompatibility Testing 1975,* Kissmeyer-Nielsen, F., Ed., Munksgaard, Copenhagen, 1975, 809.
20. **Tsuji, K., Nose, Y., Hoshino, K., Inouye, H., Ito, M., Ito, Y., Yasuda, N., and Tamaoki, N.,** Comparative study between gastric cancer patient and normal healthy control against HLA antisera, in *HLA and Disease,* INSERM, Paris, 1976, 237.
21. **Tsuji, K., Nose, Y., Shibuya, K., Hoshino, K., Inouye, H., Ito, M., and Yoshida, T.,** Ia antibody in thyroiditis, myasthenia gravis, and gastric cancer, in *HLA and Disease,* INSERM, Paris, 1976, 284.
22. **Braun, W.E.,** unpublished observation.
23. **Ryder, L.P. and Svejgaard, A.,** Associations Between HLA and Disease: Report from the HLA and Disease Registry of Copenhagen, HLA and Disease Registry, Copenhagen, 1976.
24. **Kourilsky, F.M., Dausset, J., Feingold, N., Dupuy, F.M., and Bernard, J.,** Etude de la repartition des antigens leucocytaires chez des malades atteints de leucemie aigue en remission, in *Advances in Transplantation,* Dausset, J., Hamburger, J., and Mathe, G., Eds., Williams & Wilkins, Baltimore, 1967, 515.
25. **Rogentine, G.N., Yankee, R.A., Gart, J.J., Nam, J., and Trapani, R.J.,** HL-A antigens and disease, *J. Clin. Invest.,* 51, 2420, 1972.
26. **Dausset, J. and Hors, J.,** Some contributions of the HL-A complex to the genetics of human disease, *Transplant, Rev.,* 22, 44, 1975.
27. **Albert, E.D., Mickey, M.R., and Terasaki, P.I.,** Genetics of four new HL-A specificities in the Caucasian and Negro populations, *Transplant. Proc.,* 3, 95, 1971.
28. **Pegrum, G.D., Balfour, I.C., Evans, C.A., and Middleton, V.L.,** HL-A antigens on leukaemic cells, *Br. J. Haematol.,* 19, 493, 1970.
29. **Rogentine, G.N., Trapani, R.J., Yankee, R.A., and Henderson, E.S.,** HL-A antigens and acute lymphocytic leukemia: the nature of the HL-A2 association, *Tissue Antigens,* 3, 470, 1973.

30. **Lawler, S.D., Klouda, P.T., Smith, P.G., Till, M.M., and Hardisty, R.M.**, Survival and the HL-A system in acute lymphoblastic leukaemia, *Br. Med. J.*, 1, 547, 1974.
31. **Johnson, A.H., Ward, F.E., Amos, D.B., Leikin, S., and Rogentine, N.**, HLA and acute lymphocytic leukemia, in *HLA and Disease,* INSERM, Paris, 1976, 227.
32. **Billing, R.J., Terasaki, P.I., Honig, R., and Peterson, P.**, The absence of B-cell antigen B2 from leukaemia cells and lymphoblastoid cell lines, *Lancet,* 1, 1365, 1976.
33. **Benbunan, M., Saglier, M., Owens, A., Husson, N., Bussel, A., Reboul, M., Dastot, H., and Czazar, E.**, Acute myelocytic leukemia (AML), chronic myeloid leukemia and HLA antigens, in *HLA and Disease,* INSERM, Paris, 1976, 217.
34. **Oliver, R.T.D., Klouda, P., and Lawler, S.**, HL-A associated resistance factors and myelogenous leukaemia, in *HLA and Disease,* INSERM, Paris, 1976, 231.
35. **Drew, S.I., Bergh, O.J., and Terasaki, P.I.**, Human granulocyte specific antigens and myeloid leukemia, in *HLA and Disease,* INSERM, Paris, 1976, 224.
36. **Bodmer, J.G., Bodmer, W.F., Pickbourne, P., Degos, L., Dausset, J., and Dick, H.M.**, Combined analysis of three studies of patients with Burkitt's lymphoma, *Tissue Antigens,* 5, 63, 1975.
37. **Amiel, J.L.**, Study of the leukocyte phenotypes in Hodgkin's disease, in *Histocompatibility Testing 1967,* Curtoni, E.S., Mattiuz, P.L., and Tosi, R.M., Eds., Williams & Wilkins, Baltimore, 1967, 79.
38. **Zervas, J.D., Israels, M.C., and Delamore, I.W.**, Leukocyte phenotypes in Hodgkin's disease, *Lancet,* 2, 634, 1970.
39. **Forbes, J.F. and Morris, P.J.**, Leukocyte antigens in Hodgkin's disease, *Lancet,* 2, 849, 1970.
40. **Thorsby, E., Engeset, A., and Lie, S.O.**, HL-A antigen and susceptibility to disease: a study of patients with acute lymphoblastic leukemia, Hodgkin's disease, and childhood asthma, *Tissue Antigens,* 1, 147, 1971.
41. **Falk, J. and Osoba, D.**, HL-A antigens and survival in Hodgkin's disease, *Lancet,* 2, 1118, 1971.
42. **Jeannet, M. and Magnin, C.**, HL-A antigens in haematological malignant disease, *Eur. J. Clin. Invest.,* 2, 39, 1971.
43. **Dick, F.R., Fortuny, I., Theologides, A., Greally, J., Wood, N., and Yunis, E.J.**, HL-A and lymphoid tumors, *Cancer Res.,* 32, 2608, 1972.
44. **Bertrams, J., Juwert, E., Bohme, U., Reis, H.E., Gallmeier, W.M., Werner, O., and Schmidt, C.G.**, HL-A antigens in Hodgkin's disease and multiple myeloma. Increased frequency of W18 in both diseases, *Tissue Antigens,* 2, 41, 1972.
45. **Forbes, J.F. and Morris, P.J.**, Analysis of HL-A antigens in patients with Hodgkin's disease and their families, *J. Clin. Invest.,* 51, 1156, 1972.
46. **Svejgaard, A., Platz, P., Ryder, L.P., Staub Nielsen, L., and Thomsen, M.**, HL-A disease association, *Transplant. Rev.,* 22, 3, 1975.
47. **Kissmeyer-Nielsen, F., Jensen, K.B., Ferrara, G.B., Kjerbye, K.E., and Svejgaard, A.**, HL-A phenotypes in Hodgkin's disease. Preliminary report, *Transplant. Proc.,* 3, 1287, 1971.
48. **Morris, P.J., Lawler, S., and Oliver, R.T., II**, HL-A and Hodgkin's disease, in *Histocompatibility Testing 1972,* Dausset, J. and Colombani, J., Eds., Munksgaard, Copenhagen, 1973, 669.
49. **Miller, W.V.**, HL-A antigens and hematologic malignancy, *Arch. Intern. Med.,* 133, 397, 1974.
50. **Kissmeyer-Nielsen, F., Kjerbye, K.E., and Lamm, L.U.**, HLA and Hodgkin's disease. III. A prospective study, *Transplant. Rev.,* 22, 168, 1975.
51. **Bjorkholm. M., Holm, G., Johansson, B., Mellstedt, H., and Moller, E.**, A prospective study of HL-A antigen phenotypes and lymphocyte abnormalities in Hodgkin's disease, *Tissue Antigens,* 6, 247, 1975.
52. **Osoba, D. and Falk, J.**, Genes may indicate prognosis in lymphoma, cited in Medical News, *JAMA,* 235, 2808, 1976.
53. **Cordon, A.L.**, HL-A and malignant melanoma, *Lancet,* 1, 938, 1973.
54. **VanWijck, R. and Bouillenne, C.**, HL-A antigen and susceptibility to malignant melanoma, *Transplantation,* 16, 371, 1973.
55. **Clark, D.A., Necheles, T., Nathanson, S., and Silverman, E.**, Apparent HL-A5 deficiency in malignant melanoma, *Transplantation,* 15, 326, 1973.
56. **Lamm, L.U., Kissmeyer-Nielsen, F., Kjerbye, K.E., Mogensen, B., and Petersen, N.C.**, HL-A and ABO antigens and malignant melanoma, *Cancer* (Brussels), 33, 1458, 1974.
57. **Singal, D.P., Bent, P.B., McCulloch, P.B., Blajchman, M.A., and MacLaren, R.G.**, HL-A antigens in malignant melanoma, *Transplantation,* 18, 186, 1974.
58. **Bergholtz, B., Klepp, O., Kaakinen, A., and Thorsby, E.**, HLA antigens in malignant melanoma, in *HLA and Disease,* INSERM, Paris, 1976, 218.
59. **Smith, G., Walford, R.L., Fishkin, B., Carter, P.K., and Tanaka, K.**, HL-A phenotypes, immunoglobulins and K and L chains in multiple myeloma, *Tissue Antigens,* 4, 374, 1974.
60. **Mason, D.Y. and Cullen, P.**, HL-A antigen frequencies in myeloma, *Tissue Antigens,* 5, 238, 1975.
61. **Jeannet, M. and Magnin, C.**, HL-A antigens in haematological malignant diseases, *Eur. J. Clin. Invest.,* 2, 30, 1976.

62. **Gallie, B.L., Dupont, B., Whitsett, C., Kitchen, F.D., Ellsworth, R.M., and Good, R.A.,** Histocompatibility typing in spontaneous regression of retinoblastoma, in *HLA and Disease,* INSERM, Paris, 1976, 225.
63. **Bertrams, J., Schildberg, P., Hopping, W., Bohme, U., and Albert, E.,** HL-A antigens in retinoblastoma, *Tissue Antigens,* 3, 78, 1973.
64. **Mittal, K.K., Kachru, R.B., and Brewer, J.I.,** The HL-A and ABO antigens in trophoblastic disease, *Tissue Antigens,* 6, 57, 1975.
65. **Lawler, S.D., Klouda, P.T., and Bagshawe, K.D.,** The HL-A system in trophoblastic neoplasia, *Lancet,* 2, 834, 1971.
66. **Bagshawe, K.D., Rawlins, G., Pike, M.C., and Lawler, S.D.,** ABO blood-groups in trophoblastic neoplasia, *Lancet,* 1, 553, 1971.
67. **Kissmeyer-Nielsen, F. and Thorsby, E.,** Human transplantation antigens, *Transplant. Rev.,* 4, 2, 1970.
68. **Seigler, H.F., Kremer, W.B., Metzgar, R.S., Ward, F.E., Haung, A.T., and Amos, D.B.,** HL-A antigenic loss in malignant transformation, *J. Natl. Cancer Inst.,* 46, 577, 1971.
69. **Bertrams, J., Kuwert, E., Gallmeier, W.M., Reis, H.E., and Schmidt, C.G.,** Transient lymphocyte HL-A antigen "loss" in a case of irradiated M. Hodgkin, *Tissue Antigens,* 1, 105, 1971.
70. **Old, L.J., Stockert, E., Boyse, E.A., and Kim, J.H.,** Antigenic modulation. Loss of TL antigen from cells exposed to TL antibody. Study of the phenomenon in vitro, *J. Exp. Med.,* 127, 523, 1968.

X. NEUROLOGY AND PSYCHIATRY

A. Multiple Sclerosis (MS)

The first antigen shown to occur with increased frequency in this disease is B7 (combined uncorrected $P < 7.6 \times 10^{-10}$) with a relative risk (RR) of 1.55 and 95% confidence interval of 1.36 to 1.81.[1] The major studies in this disease are shown in Table 17.

In 1972, several studies described the possible association of multiple sclerosis (MS) with an HLA antigen. Naito reported on 94 cases and found that the most significant difference was a frequency for A3 of 40.4% of patients compared to 23.5% of controls.[2] Bertrams reported on 393 German patients with MS and found that the A3 antigen occurred in 35.6% of patients compared to 27.1% of controls, an insignificant difference.[3] The only significantly increased antigen in that very large study was W5, now called Bw35, which occurred in 22.6% of patients compared to 11% of controls. Jersild reported a significant increase for only B7 in 107 Danish patients.[4] In the Italian population, Cazzullo found a significant increase in the frequency of only A9 to 41% in 74 MS patients compared to 27% in controls.[5] Thus, in the initial four studies of HLA and disease association in MS, different antigens were selected by four different studies.

More significant information which avoided the early confusion surrounding the A- and B-series antigens was presented later by Jersild when he reported that 19 of 28 MS patients (70%) had a D-series antigen identified then as LD-7a and now known as Dw2, which is in linkage disequilibrium with B7.[6] This frequency was much higher than the 16% observed in healthy individuals without B7 and even higher than the frequency of 56% found in healthy individuals who had'Dw2 in linkage disequilibrium with B7 because 100% of the 13 patients with MS who had B7 were also Dw2-positive. In examining the relationship that the presence of this antigen had to the clinical course, these authors noted that those patients lacking Dw2 progressed more slowly, whereas those who carried the Dw2 antigen could be divided into slowly progressing and rapidly progressing groups that were not additionally influenced by the presence or absence of B7. Thus, the Dw2 antigen not only appeared to increase the susceptibitity to develop the disease but also seemed to enhance the possibilities of a more aggressive course of the disease.

TABLE 17

Multiple Sclerosis in Caucasians and American Blacks

HLA Ag	Controls with Ag (%)	Patients with Ag (%)	Patients with Ag (n)	Patients total (n)	Relative risk	Ref.
			Caucasians			
B7	23.5	27.7	26	94	1.2	2
	21.0	39.3	22	56	2.4	22
	21.1	28.1	9	32	1.5	29
	26.8	39.7	83	209	1.8	28
	26.0	43.2	54	125	2.2	25
	30.2	46.0	21	46	1.9	14
Dw2	18.0	60.0	29	48	6.7	28
	15.4	72.2	13	18	14.3	28
	14.0	51.4	57	111	6.5	7
	18.3	52.0	172	330	5.1	8
	30.2	68.0	25	37	4.8	14
Pi	19.6	43.6	17	39	3.2	11
			American Blacks			
B7	29.4	35.5	11	31	1.3	9
Dw2	—	35.5	11	31	—	9

HLA-D-series antigens Dw1, Dw2, Dw3, and a more recently defined determinant, EI, were tested in 111 MS patients from West Germany.[7] Only the frequency of Dw2 was found to be significantly elevated, occurring in 51.4% of the patients compared to 14% of 150 controls. Dw2 was also found to be associated with B7 and B18 in 72.5% of the B7- or B18-positive MS patients, which was significantly different from the appropriate controls ($P < 0.0001$). Dw3, which was not significantly different from the control group, occurred in 23.8% of the MS patients who lacked B8, which was significantly higher than in normal controls ($P < 0.05$), suggesting a possible disturbance of the gametic association between B8 and Dw3 in MS patients.[7]

The largest study thus far of MS disclosed that the Dw2 antigen occurred in 52% of 330 Caucasian patients compared to 18% of 136 normal Caucasian controls.[8] The HLA-A3-B7-Dw2 haplotype was found in 17% of patients compared to only 7% of controls. Neither A3 nor B7 alone was significantly increased, but the joint occurrence of A3-Dw2 (20% compared to 7% of controls) and B7-Dw2 (30% compared to 13% of controls) was strongly associated with MS. From this study, it was apparent that the D-locus antigen Dw2 was the single antigen most strongly related to MS.

Of 31 Black Americans with MS, 11 had Dw2 ($P < 0.001$) which was the only significantly altered antigen.[9]

Using serologic techniques, several investigators have now discovered antisera that possibly detect an Ia-like antigen which has a very high association with MS. The first to report such an antigen was Winchester, who found that 100% of 25 patients with MS reacted with a pregnancy serum #770 which detected an antigen called Ag 7a by immunofluorescence.[10] Even more impressive was the fact that within this group of 25 patients with MS, there were 16 who did not have the Dw2 antigen but who all had the Ag 7a serologically defined antigen. Serum Pi of Mempel, Serum SOW of Thorsby, and another serum of Wernet similarly seemed to identify MS patients with a 44% to better than 90% accuracy.[11-14] Terasaki tested 56 MS patients for six new B-cell antigens and found that 84% had the group 4 specificity compared to 32% of normals.[15]

In the family studies of Jersild, further evidence was provided in support of MS being a polygenic disease.[6] HLA-identical siblings did not always show concordance for the disease. Although the Dw2 antigen was inherited and linked to a specific HLA haplotype in some families, this occurred even when the patients were B7-negative. The fact that Dw2 was present in unaffected as well as affected individuals could be explained by incomplete penetrance of MS. But the finding that 30% of unrelated MS patients lacked Dw2 supported the contention that MS is a polygenic disease.

Other family studies in MS patients were inconclusive and showed only a weak increase in the A3-B7-Dw2 haplotype. In an analysis of eight families, a particular haplotype was associated with MS within each family, but from family to family, the haplotype involved with MS differed.[16] In an analysis of 28 pairs of siblings in Western Germany, both of whom had MS, an analysis for joint segregation with HLA showed that 24 of 28 pairs shared 1 haplotype, a finding compatible with joint segregation of MS and HLA. [17] It should be noted that these data are preliminary and a combined analysis is not possible at this time.

The finding that lymphocytes from MS patients did not release leukocyte migration inhibitory factor when exposed to measles and other myxoviruses in vitro, although lymphocytes from most normal individuals did, suggested that MS patients lack cellular immunity mediated by T-cell responses to those myxoviruses while maintaining a normal or increased B cell-mediated antibody response to these same viruses.[18] This information would suggest that the defect in MS involved a defect in T cells that made the MS patients incapable of eradicating the myxovirus and that in some way was associated with the Dw2 histocompatibility antigen. Others have reported lower responses to PHA and Con-A in MS patients who are A3- and B7-positive than in those who are A3-B7-negative,[19] further incriminating the T cell.

As additional evidence of a T-cell deficiency in MS, it has been shown that only T lymphocytes have measle receptors and that patients with MS have a slightly increased number of B cells in the peripheral blood, whereas T lymphocytes are reduced.[20] However, others have recently demonstrated that measle virus can replicate in both human T and B lymphocytes as well as monocytes. [20] Nevertheless, numerous studies have shown that patients with MS have slightly but consistently elevated serum measle antibody titers as compared to controls. Jersild has reported that antibody titers to measle virus greater than 1:256 were found significantly more often among 57 MS patients carrying the HLA antigens A3, B7, or B18 than in 45 MS patients lacking these antigens.[21] In exploring the fact that patients with MS and the A3 antigen had elevated measle antibody titers, Arnason found that healthy individuals with A3 had increased measle antibody titers as well.[22] This suggested that the elevated measles antibody titer may be a reflection of the histocompatibility antigen A3 rather than of MS.

A correlation with various antibody titers in 110 patients with MS and 85 controls was sought in relationship to the presence of A3, B7, A2, B12, Bw35, B15, and B18.[23] In MS patients, a significant increase was found in antibodies against several measle antigens but not against the Sendae or mumps virus. The antibody titer of these viral antigens could not be correlated with any HLA antigens, including the haplotype A3-B7. Forty other MS patients were studied for both humoral (antibody titer) and cell-mediated (lymphocyte transformation) immunity to five viruses (measles, rubella, periinfluenza 3, mumps, and herpes simplex).[24] No significant differences in the viral responses were found, and no correlation between the HLA typing for A- and B-series antigens or for the Dw2 specificity and viral immunity was detected.

However, other studies of measle antibodies in MS patients conducted by Paty showed that patients with A3, B7, and B18 had elevated mean complement fixing measle titers compared to patients not carrying these antigens.[19] The clinical course of the A3-, B7- and B18-negative male patients was more severe and rapid than in other patients. Although this was a restricted group, the results were somewhat at variance with Jersild's finding that Dw2-negative patients had a more slowly progressive disease and B7 did not further affect the course.[6]

These results, though certainly not uniform, suggested that the increased antibody level and reduced mitogen responsiveness in MS patients may reflect a T-cell deficiency, possibly linked to the HLA system, which appears to be important both in terms of susceptibility to the disease and the clinical expression of the disease.

In MS, there may be local production of measle antibody in the central nervous system as reflected in elevated levels of antibody in the CSF.[20] However, once again, the data are conflicting. Studies from Australia[24] and Norway[14] showed no correlation between antibody titers, cerebrospinal fluid, IgG levels, and HLA. Thorsby's study from Norway demonstrated that oligoclonal IgG was present in the cerebrospinal fluid of 19 of 20 Dw2-positive patients and 6 of 7 Dw2-negative patients. Similarly, 13 of 18 Dw2-positive and 2 of 5 Dw2-negative patients had a reduced serum-to-CSF ratio of measle antibodies, while 3 of 13 Dw2-positive and 0 of 5 Dw2-negative patients had reduced serum-to-CSF ratios of rubella antibody. Using a more elaborate analysis, Oger found a high intrathecal secretion of IgG in individuals with B7 as well as those with B12 and Bw35, which suggested to these authors that perhaps two different immune response genes were responsible for the cerebrospinal fluid abnormalities in MS.[25]

At present, there is no satisfactory explanation for the discrepancy between the studies showing an increased antibody level in some patients with MS and no increase in others. However, the depression of cellular immunity as tested by mitogen stimulation does seem to be confirmed thus far, although the capacity of MS patients for measles-specific MIF production has had conflicting results.[20]

The complement profiles and properdin level of patients with MS were examined in 75 patients.[26] Low C3 levels were found in 12 cases, low C3 and factor B levels in 15 cases, low factor B levels in 14 cases, and no complement alteration in 34 cases. Those with low C3 levels alone or with low factor B levels showed a significant association with B18 (25.9% compared to 7.6% of controls), largely due to the subgroup with low properdin levels in which the frequency of B18 was 33.3%. In contrast, the normal complementemic group of 34 patients was strongly associated with B7.

Because of the association of Dw2 with C2 (see Complement Deficiencies [Chapter 7, Section III]) Bertrams examined the association between C2 levels and MS.[27] He found that patients with MS had levels of C2 that were generally lower than those of the normal population but had a bimodal distribution. The Dw2-positive MS patients had lower levels of C2 and the Dw2-negative MS individuals had higher levels which partially overlapped with the lower range of normals.

In general, MS has HLA associations of increasing strength as one progresses from A3, to B7, to Dw2, and finally to the serologically defined Ia antigens. Families with MS revealed the following facts:

1. MS is not a simple genetic disease showing Mendelian inheritance patterns.
2. The probands do not always possess what has been called an MS chromosome.
3. In some families, the MS susceptibility segregates independent of HLA.
4. The associated C2 deficiencies found in MS cannot be conclusively identified as either primary or secondary from the information available.

Finally, as a counterpart to the increased frequency of B7 in MS and its A-series antigen in linkage disequilibrium, namely, A3, there was the observation that A2 and B12 are decreased, possibly to an extent greater than could be accounted for by the relative increase in A3 and B7. This has raised the possibility that the decreased frequency of 2 and 12 may be linked to a hypothetical gene for resistance to MS.[28]

TABLE 18

Optic Neuritis

HLA Ag	Controls with Ag (%)	Patients with Ag (%)	Patients with Ag (n)	Patients total (n)	Relative risk	Ref.
B7	26.7	39.0	21	54	1.7	31
Dw2	18.0	50.0	27	54	4.5	31

B. Optic Neuritis

A study by Platz of 54 patients who had only optic neuritis (Table 18) at the beginning of an initial observation period lasting 6 months to 6 years demonstrated an increased frequency of A3 and B7 very similar to that found in his patients with MS.[30] For example, A3 occurred in 33% of 54 patients with optic neuritis and in 35% of those with MS compared to 26.9% of controls. B7 occurred with a frequency of 39% in both optic neuritis and MS compared to 26.7% in controls. The D-locus antigen, then called LD-7a and now known as Dw2, occurred in 50% of 54 patients with optic neuritis, 60% of 233 with MS and only 18% of 132 controls. The Dw2 antigen was also found in 5 of 11 patients who developed MS during the observation period and in 12 of the 25 patients who had optic neuritis with oligoclonal IgG. These authors were unable to confirm the results of Arnason, who found only an increased frequency of the A1-A3 phenotype in 20 patients with optic neuritis.[22]

C. Subacute Sclerosing Panencephalitis (SSPE)

Because the pathogenesis of SSPE may be related to impaired cell-mediated immunity against the measle virus, a possible association has been sought with HLA antigens. However, 32 patients with SSPE showed no significant deviation in the HLA-A and B antigens tested.[31]

D. Amyotrophic Lateral Sclerosis (ALS)

In two studies of ALS, 24[32] and 25[33] patients disclosed no significant association between this disease and any of the A-, B-, C-, and D-series antigens tested.

E. Myasthenia Gravis

The results within this disease entity are quite consistent and show a significant increase in the frequency of B8 (combined uncorrected $P < 10^{-10}$) with a relative risk (RR) of 4.40 and 95% confidence interval of 3.33 to 5.82.[1] The major studies in this disease are shown in Table 19. The other associations with A1 and Dw3 were felt to be due to linkage disequilibrium, but neither of these antigens was as high in its frequency association with this disease as was B8.

In the two major clinical groups of patients with myasthenia gravis, namely those with a thymoma and hyperplasia, B8 was most strikingly elevated in those who had hyperplasia as well as no autoantibody against striated muscle, an early onset of the disease under the age of 40 years, and were female. In contrast, B8 did not seem to be increased in those who had a thymoma, autoantibody, late onset of disease, or were male. In a Finnish study in which the frequency of B8 in controls was 18%, the B8 frequency of 159 myasthenics increased to 72% in females with the onset of the disease before the age of 35 and to 74% in patients with thymic hyperplasia.[34] This study also included typing for the Dw3 antigen, which was found to be present in 63% of myasthenics compared to 50% of controls who also had B8. The occurrence of B8 in myasthenics who had associated immunologic and thyroid disorders was actually lower

TABLE 19

Myasthenia Gravis in Caucasians

HLA Ag	Controls with Ag (%)	Patients with Ag (%)	Patients with Ag (n)	Patients total (n)	Relative risk	Ref.
	18.1	48.6	17	35	4.2	41
	31.2	65.4	17	26	4.1	40
	20.0	47.6	20	42	3.6	42
	19.3	59.0	59	100	6.0	1
	17.8	37.5	21	56	2.7	3
	18.1	—	—	159	4.3	34
Females 36 years	—	—	—	—	12.2	34

(44%) than those myasthenics with no additional disease (66%), suggesting that one of these disorders might facilitate the development of myasthenia when B8 is absent. B8 was found in eight of ten familial myasthenics and in 49% of 127 healthy relatives of myasthenics.[34]

Further support for the combined HLA and sex distribution in myasthenics was offered by Fritze, who found that 50% (19/37) of the females had B8 compared to only 2/19 males among 56 Caucasians with myasthenia gravis.[35] They suggested that because females are known to have approximately twice the incidence of myasthenia gravis as males, the sex chromosome as well as the HLA loci might contribute to the genetic determination of myasthenia gravis. This concept conforms to the polygenic inheritance patterns involving sex chromosomes and non-sex-linked histocompatibility systems which affect disease incidence in mice.[36] Therefore, the relative risk of myasthenia gravis in Caucasians differs significantly according to the sex of the individual, so that for a male with B8 it is approximately 4.3 whereas for a female it is 12.2.[34]

Studies by Kaakinen and Möller showed an increased frequency of Dw3 in patients with myasthenia gravis.[37,38] The interesting point made by Möller's study was that myasthenia gravis was more closely related to B8 than it was to the Dw3 specificity.[38] Moreover, the association with B8 or Dw3 was more pronounced in females. Specifically, 37 females were studied of whom 26 had B8 and only 15 had Dw3, whereas of 7 males with the disease, only 2 had B8 and 1 had Dw3. Although the male group was small, there was the suggestion that the antigen associations were more striking in females and that the B8 association was stronger than that for Dw3.

Thus, myasthenia gravis may be different from the other B8-associated diseases (thyrotoxicosis in Caucasians, celiac disease, JDM, Addison's disease, chronic active hepatitis, and dermatitis herpetiformis, all of which may have an autoimmune component, too) in that its association is stronger with the B-series antigen B8 than it is with the D-series antigen Dw3.

However, the B8 association was lost in 65 Japanese patients with myasthenia gravis (44 females and 21 males). In Japanese myasthenics, three different antigens were significantly increased, namely, A9, B12, and Bw35. The most frequent haplotype in family studies was A9-B5. When one examined the subgroups according to the type of thymic disease that was present, it was noted that B5 was more common in those myasthenics who had a thymoma, whereas B12 was more frequent in young females with hyperplasia.[39]

In the few reported patients, homozygosity for A1 and B8 was not associated with a more severe congenital myasthenia and identical twins apparently with B8 were discordant for the disease.[40]

F. Motor Neuron Disease

In a preliminary report from Scotland, 32 patients (26 males and 6 females) suffering from motor neuron disease described as a selective degeneration of neurons involving the nuclei of the brain stem, anterior horn cells, and both cortical spinal pathways, and having an hereditary component, showed a surprisingly high frequency of the A-series antigens A2 and A28 that occurred in 78.2% of the patients.[43] The small sample size of this particular study must be considered as one of the possible causes of this unusual finding.

G. Vitamin B12 Neuromyelopathy (including Pernicious Anemia)

In one study, 12 cases of pernicious anemia with neuromyelopathy were compared with 54 patients having pernicious anemia without neurologic damage.[44] An increased frequency of the phenotype A2-B12 was found in those with neuromyelopathy (58% compared to 6%). Pernicious anemia is discussed under Gastroenterology (Chapter 7, Section VII).

H. The Lennox-Gastaut Syndrome

This disease, which consists of mental retardation and minor motor seizures from a wide variety of causes, has been found to have an increased frequency of B7 and of an unspecified D antigen in 15 unrelated children.[45]

I. Paralytic Dementia Secondary to Syphilis

It was found that 59 American Caucasians and 35 American Blacks institutionalized with the late stage of neurosyphilis presenting as paralytic dementia and confirmed by serologic testing had, in the Caucasian patients only, an increased frequency of B18 (14/59 patients — 24%). No other antigen associations were found and no deviations were found in the American Black patients.[46]

J. St. Louis Encephalitis

Following an epidemic of St. Louis encephalitis in Greenville, Mississippi in which approximately 100 individuals from a population of about 50,000 had symptomatic infection, a study was performed to examine the HLA frequencies among the affected individuals and an appropriate control group. A total of 26 American Blacks and 9 Caucasian patients were studied.[47] A decreased frequency of Bw35 was found in the Black patients when compared to Black controls (30% compared to 54%). No other significant differences in HLA antigens were found. A small number of patients were studied for the D-series antigens Dw1 and Dw2, but no predominant antigen was found in either the Blacks or the Caucasians. Thus, symptomatic St. Louis encephalitis due to group B Togavirus does not seem to be related to any specific HLA antigen and in particular, B7 and Dw2 were not increased among these patients.

K. Paralytic Poliomyelitis

Despite suggestive evidence that B7 was found with an increased frequency in an early study of paralytic poliomyelitis, [48] subsequent studies have failed to confirm this.[29,49,50] Although he found no increase in B7, the study by Zander did find a significant decrease in the frequency of B8 (6% compared to 19% in a control population of 5,046 with a P value < 0.013).[50] However, the Dw3 specificity known to be in linkage disequilibrium with B8 was not diminished. A result that in part supported the initial study of Pietsch was the finding that Dw2, known to be in linkage disequilibrium with B7, was increased (28% compared to 14% in a control population of 307; $P < 0.035$).[50] The only conclusions that can be drawn from these conflicting data are that there is only weak evidence that the B7 and Dw2 antigens might be increased in paralytic poliomyelitis and that the B8 antigen alone might be decreased without any associated decline in Dw3.

L. Spinocerebellar Ataxias

One of this group of inherited spinocerebellar ataxias, the late-onset, dominantly inherited Marie's ataxia, was studied in a single Japanese family. The father who died of the disease and three female children who received his A9-B5 haplotype all had the disease, whereas two children, a boy and a girl, who received the other paternal haplotype, A11-Bw40, were healthy.[51] This suggestive evidence that the ataxia-gene locus might be on the sixth chromosome has recently been solidly established by linkage analysis on 19 members of a kindred with dominantly inherited spinocerebellar ataxia.[52] In this family, the disease occurred with the A3-B14 haplotype.

M. Migraine Headache

In a study of 21 unrelated patients with carefully diagnosed migraine, there was an increased frequency of B12 (68% compared to 29.3% in healthy controls)[53] However, B12 did not occur more frequently in the migraine headache group than in those with cluster headaches. In one family study, migraine seemed to be associated with the haplotype A2-B12.

N. Schizophrenia and Manic Depressive Disorders

In 38 schizophrenics and 18 manic depressive patients, HLA antigen frequencies analyzed by discriminant analysis showed that in the schizophrenic patients, there were no patients with A11 compared to 13% in controls ($P < 0.01$), only 8% with B7 compared to 24% in the controls ($P < 0.05$), and 3% with B8 compared to 20% in controls ($P < 0.02$). B17 was the only antigen found more frequently (18% of patients compared to 7% of controls — $P < 0.05$).[54] The higher frequency carried over to the manic depressives where 28% had B17 compared to 6% of controls ($P < 0.01$).

Quite different results were obtained in Czechoslovakia in a study of 148 male schizophrenic patients and 1200 healthy controls.[55] In this study, A28 was found in 18.9% of patients compared to 6.3% of the control group (corrected $P < 0.01$). The relative risk for such individuals was calculated to be 3.45.

Subtypes of schizophrenia were studied by Mercier.[56] Increased frequencies of A9 (46.6% compared to 24% of controls) and B5 (40% compared to 13.2% of controls) were found in 15 paranoid schizophrenics. A tendency for A1 to be increased was also noted, but this was equally divided between five hebephrenic and five paranoid patients. These results tended to confirm the earlier results of Cazzullo,[57] although the number of patients studied was small.

Because A9 and A28 are frequently present on haplotypes with Cw4, the 40% frequency of Cw4 in 40 patients with paranoid schizophrenia (compared to 15.3% in 1200 controls) seemed to offer a unifying link for at least these two different A-series antigens.[58]

O. Ataxia Telangiectasia

This disease is discussed under Immunodeficiency Diseases (Chapter 7, Section III).

REFERENCES

1. **Ryder, L.P. and Svejgaard, A.,** Associations Between HLA and Disease: Report from the HLA and Disease Registry of Copenhagen, 1976.

2. **Naito, S., Namerow, N., Mickey, M.R., and Terasaki, P.I.**, Multiple sclerosis: association with HL-A3, *Tissue Antigens*,2, 1, 1972.
3. **Bertrams, J., Kuwert, E., and Liedtke, U.**, HL-A antigens and multiple sclerosis, *Tissue Antigens*, 2, 405, 1972.
4. **Jersild, C., Svejgaard, A., and Fog, T.**, HL-A antigens and multiple sclerosis, *Lancet*, 1, 1240, 1972.
5. **Cazzullo, C.L. and Smeraldi, E.**, HL-A antigens and multiple sclerosis, *Lancet*, 2, 429, 1972.
6. **Jersild, C., Hansen, G.S., Svejgaard, A., Fog, T., Thomsen, M., and Dupont, B.**, Histocompatibility determinants in multiple sclerosis, with special reference to clinical course, *Lancet*, 2, 1221, 1973.
7. **Grosse-Wilde, H., Bertrams, J., Netzel, B., and Kuwert, E.K.**, Frequency of four HLA-D alleles in multiple sclerosis patients, in *HLA and Disease*, INSERM, Paris, 1976, 67.
8. **Opelz, G., Terasaki, P., Myers, L., Ellison, G., Ebers, G., Zabriskie, J., Weiner, H., Kempe, H., and Sibley, W.**, The association of HLA antigens A3, B7, and Dw2 with 330 multiple sclerosis patients in the United States, *Tissue Antigens*, 9, 54, 1977.
9. **Dupont, B., Lisak, R.P., Jersild, C., Hansen, J.A., Silberberg, D.H., Whitsett, C., Zweiman, B., and Ciongoli, K.**, HLA antigens in Black American patients with multiple sclerosis, in *HLA and Disease*, INSERM, Paris, 1976, 65.
10. **Winchester, R.J., Ebers, G., Fu, S.M., Espinosa, L., Zabriskie, J., Kunkel, H.G.**, B-cell alloantigen Ag 7a in multiple sclerosis, *Lancet*, 2, 814, 1975.
11. **Mempel, W., Grosse-Wilde, H., and Albert, E.D.**, cited in **Jersild, C., Dupont, B., Fog, T., Platz, P., and Svejgaard, A.**, Histocompatibility determinants in multiple sclerosis, *Transplant. Rev.*, 22, 148, 1975.
12. **Thorsby, E.**, personal communication.
13. **Wernet, P.**, personal communication.
14. **Thorsby, E., Helgesen, A., Solheim, B.G., and Vandvik, B.**, HLA antigens in multiple sclerosis, in *HLA and Disease*, INSERM, Paris, 1976, 84.
15. **Terasaki, P.I., Park, M.S., Opelz, G., and Ting, A.**, Multiple sclerosis and high incidence of a B lymphocyte antigen, *Science*, 193, 1245, 1976.
16. **Olsson, J.E., Link, H., and Möller, E.**, HLA antigens in families with two or more members with multiple sclerosis, in *HLA and Disease*, INSERM, Paris, 1976, 75.
17. **Zander, H., Kuntz, B., Scholz, S., and Albert, E.D.**, Analysis for joint segregation of HLA and multiple sclerosis in families, in *HLA and Disease*, INSERM, Paris, 1976, 91.
18. **Utermohlen, V. and Zabriskie, J.B.**, A suppression of cellular immunity in patients with multiple sclerosis, *J. Exp. Med.*, 138, 1591, 1973.
19. **Paty, D.W., Cousin, H.K., Stiller, C.R., Furesz, J., and Boucher, D.W.**, HLA-A & B in multiple sclerosis; relationship to measles antibody, mitogen responsiveness, and clinical course, in *HLA and Disease*, INSERM, Paris, 1976, 77.
20. **Hirsch, M.S.**, Virus markers in multiple sclerosis, *N. Engl. J. Med.*, 294, 1457, 1976.
21. **Jersild, C., Ammitzboll, T., Clausen, J., and Fog, T.**, Association between HL-A antigens and measles antibody in multiple sclerosis, *Lancet*, 1, 151, 1973.
22. **Arnason, B.G., Fuller, T.C., Lehrich, J.R., and Wray, S.H.**, Histocompatibility types and measles antibodies in multiple sclerosis and optic neuritis, *J. Neurol. Sci.*, 22, 419, 1974.
23. **Kuwert, E.K., and Hoher, P.G.**, Lack of association between HLA-system and antibodies to whole virus and virus-subunits of members of the paramyxovirus group in MS-patients and controls, in *HLA and Disease*, INSERM, Paris, 1976, 70.
24. **Stewart, G.J., Basten, A., Guinan, J., and Bashir, H.**, HLA-Dw2, viral immunity and multiple sclerosis, in *HLA and Disease*, INSERM, Paris, 1976, 83.
25. **Oger, J., Chales, G., Bansard, J.Y., Sabouraud, O., Raingeard, P., and Kerbaol, M.**, Relation between the intrathecal secretion of IgG and HLA-B phenotypes in multiple sclerosis, in *HLA and Disease*, INSERM, Paris, 1976, 73.
26. **Trouillas, P., Betuel, H., and Devic, M.**, Hypocomplementaemic and normocomplementaemic multiple sclerosis. Correlations with specific HLA antigens (B18 and B7), in *HLA and Disease*, INSERM, Paris, 1976, 85.
27. **Bertrams, J.**, personal communication.
28. **Jersild, C., Dupont, B., Fog, T., Platz, P., and Svejgaard, A.**, Histocompatibility determinants in multiple sclerosis, *Transplant. Rev.*, 22, 148, 1975.
29. **Dausset, J. and Hors, J.**, Some contributions of the HL-A complex to the genetics of human diseases, *Transplant. Rev.*, 22, 44, 1975.
30. **Platz, P., Ryder, L.P., Staub-Nielsen, L., Svejgaard, A., Thomsen, M., and Wolheim, M.S.**, HL-A and idiopathic optic neuritis, *Lancet*, 1, 520, 1975.
31. **Reinert, Ph., Mannoni, P., Lebon, P., and Ponsot, G.**, HLA and subacute sclerosing panencephalitis (SSPE), in *HLA and Disease*, INSERM, PAris, 1976, 80.
32. **Bartfeld, H., Whitsett, C., and Donnenfeld, H.**, HLA antigens and amyotrophic lateral sclerosis, in *HLA and Disease*, INSERM, Paris, 1976, 61.

33. **Pedersen, L. and Platz, P.,** Tissue-typing in amyotrophic lateral sclerosis, in *HLA and Disease,* INSERM, Paris, 1976, 78.
34. **Pirskanen, R. and Tiilikainen, A.,** Myasthenia gravis and HLA, in *HLA and Disease,* INSERM, Paris, 1976, 79.
35. **Fritze, D., Herman, C., Smith, G., and Walford, R.,** HL-A types in myasthenia gravis, *Lancet,* 2, 211, 1973.
36. **Muhlbach, O.,** The value of experimental cancer research for the understanding of human disease, in *RNA Virus and Host Genome in Oncogenesis,* American-Elsevier, Amsterdam, 1972, 339.
37. **Kaakinen, A., Pirskanen, R., and Tiilikainen, A.,** LD antigens associated with HL-A8 and myasthenia gravis, *Tissue Antigens,* 6, 175, 1975.
38. **Möller, E., Hammarstrom, L., Smith, E., and Matell, G.,** HL-A8 and LD-8a in patients with myasthenia gravis, *Tissue Antigens,* 7, 39, 1976.
39. **Yoshida, T., Tsuchiya, M., Shimabukuro, K., Satoyoshi, E., Tamaoki, N., and Tsuji, K.,** HLA antigens and myasthenia gravis in Japan, in *HLA and Disease,* INSERM, Paris, 1976, 89.
40. **Behan, P.O., Simpson, J.A., and Dick, H.,** Immune response genes in myasthenia gravis, *Lancet,* 2, 1033, 1973.
41. **Pirskanen, R., Tiilikainen, A., and Hokkanen, E.,** Histocompatibility (HL-A) antigens associated with myasthenia gravis, *Ann. Clin. Res.,* 4, 304, 1972.
42. **Safwenberg, J., Lindblom, J.B., and Osterman, P.O.,** HL-A frequencies in patients with myasthenia gravis, *Tissue Antigens,* 3, 465, 1973.
43. **Behan, P.O., Durward, W.F., and Dick, H.,** Histocompatibility antigens associated with motorneurone disease, *Lancet,* 2, 803, 1976.
44. **Horton, M.A. and Oliver, R.T.D.,** HLA phenotype A2; B12 in vitamin B_{12} neuromyelopathy, in *HLA and Disease,* INSERM, Paris, 1976, 68.
45. **Smeraldi, E., Scorza-Smeraldi, R., Guareschi-Cazzullo, A., Cazzullo, C.L., Rugarli, C., and Canger, R.,** Immunogenetics of Lennox-Gastaut syndrome: Search for LD determinants as genetic markers of the syndrome, in *HLA and Disease,* INSERM, Paris, 1976, 82.
46. **Whitsett, C., Turner, W.J., and Dupont, B.,** HLA antigens in paralytic dementia (syphilis), in *HLA and Disease,* INSERM, Paris, 1976, 88.
47. **Whitsett, C., Lee, T., Powell, K., Dupont, B., and Lytle, V.,** HLA antigens in St. Louis encephalitis, in *HLA and Disease,* INSERM, Paris, 1976, 87.
48. **Pietsch, M.D. and Morris, P.J.,** An association of HL-A3 and HL-A7 with paralytic poliomyelitis, *Tissue Antigens,* 4, 50, 1974.
49. **Lacert, Ph., Durand, J.J., Gauardin, M., Grossiord, A., Gony, J., and Hors, J.,** HLA and poliomyelitis, in *HLA and Disease,* INSERM, Paris, 1976, 71.
50. **Zander, H., Grosse-Wilde, H., Scholz, S., Kuntz, B., Netzel, B., and Albert, E.D.,** Poliomyelitis: Analysis of HLA-A, -B, and -D alleles, in *HLA and Disease,* INSERM, Paris, 1976, 90.
51. **Yakura, H., Wakisaka, A., Fujimoto, S., and Itakura, K.,** Hereditary ataxia and HL-A genotypes, *N. Engl. J. Med.,* 291, 154, 1974.
52. **Jackson, J.J., Currier, R.D., Terasaki, P.I., and Morton, N.E.,** Spinocerebellar ataxia and HLA linkage, *N. Engl. J. Med.,* 296, 1138, 1977.
53. **Devic, M., Trouillas, P., Betuel, H., and Aimard, G.,** HLA-B12 and migraine, in *HLA and Disease,* INSERM, Paris, 1976, 299.
54. **Bennahum, D.A., Troup, G.M., Rada, R.T., Kellner, R., and Kyner, T.,** HLA antigens in schizophrenic and manic depressive mental disorders, in *HLA and Disease,* INSERM, Paris, 1976, 64.
55. **Ivanyi, D., Zemek, P., and Ivanyi, P.,** HLA antigens in schizophrenia, *Tissue Antigens,* 8, 217, 1976.
56. **Mercier, P., Kieffer, N., Julien, R., and Sutter, J.M.,** Schizophrenia: HLA-A9 and B5 antigens, in *HLA and Disease,* INSERM, Paris, 1976, 72.
57. **Cazzullo, C.L., Smeraldi, E., and Penati, G.,** The leukocyte antigenic system HL-A as a possible genetic marker of schizophrenia, *Br. J. Psychiatry,* 125, 25, 1974.
58. **Ivanyi, P., Ivanyi, D., and Zemek, P.,** HLA-Cw4 in paranoid schizophrenia, *Tissue Antigens,* 9, 41, 1977.

XI. OPHTHALMOLOGIC DISEASES

A. Acute Anterior Uveitis

The primary antigen shown to occur with increased frequency in this disease is B27

TABLE 20

Acute Anterior Uveitis

HLA Ag	Controls with Ag (%)	Patients with Ag (%)	Patients with Ag (n)	Patients total (n)	Relative risk	Ref.
			Overall			
B27	7.3	58.0	58	100	17.0	3
	8.2	55.6	51	90	13.7	4
			Without Associated Disease			
	7.3	43.3	29	67	9.64	3
	8.2	44.4	28	63	9.09	4
			With Associated Disease			
	7.3	87.9	29	33	91.61	3
	8.2	85.2	23	27	64.76	4

(combined uncorrected $P < 1.0 \times 10^{-10}$) with a relative risk (RR) of 15.39 and 95% confidence interval of 10.09 to 23.46.[1] The major studies in this disease are shown in Table 20.

Because of the association of acute anterior uveitis with ankylosing spondylitis which has such a strong B27 association, anterior uveitis was examined for its possible HLA association. B27 was identified in 52% (26/50) of uveitis patients compared to 4% (2/50) of controls.[2] Of the 26 B27-positive patients, 18 had an associated disease and 8 did not. In a follow-up study of 100 patients, B27 was present in 29 of 33 with and in 29 of 67 without an associated disease.[3] The finding of B27 in 10 of 13 women patients less than 35 years of age suggested that young women with B27 would be likely to present themselves to an ophthalmologist with uveitis alone whereas young men with B27 would present themselves to a rheumatologist because of ankylosing spondylitis or Reiter's disease sometimes complicated by uveitis.

Woodrow reported that 55.7% of 90 patients with non-granulomatous uveitis had B27 compared to 8.2% of controls, a result very similar to that of Brewerton.[4] In a study of 63 patients who had no systemic disease, 28 (44.4%) had B27, giving a relative risk of 14.4.

In Blacks with ankylosing spondylitis, iritis was significantly more frequent in those with B27.[5]

B. Primary Open Angle Glaucoma

Of 80 patients with primary open-angle glaucoma, 40 Caucasians and 40 Blacks, 50% were found to have B12 compared to 11% of controls and 49% had B7 compared to 20% of controls. [6] Either B12 or B7 was present in 88% of patients with primary open-angle glaucoma compared to 30% of the general population. In another study, Bw35 occurred in 46.9% (23 of 49) of patients compared to 22.8% of 250 controls (corrected $P < 0.012$, with a relative risk of 3.0).[7] However, B7 or B12 was found in only 18.3% of the patients in the second study.

C. Ocular Hypertension

The same authors who studied primary open-angle glaucoma also investigated 76 patients with ocular hypertension and found that 88% of those who developed glaucomatous visual field loss had either B12 or B7.[8] Of 34 patients, 14 (41%) with ocular hypertension and a gg response (after topical corticosteroids, a dramatic response with applanation pressures exceeding 31 mmHg) who had either B12 or B7 developed a glaucomatous visual-field loss whereas only 2 of 41 similar patients without either antigen developed a visual-field loss.

D. The Vogt-Koyanagi-Harada Syndrome

This syndrome of bilateral granulomatous uveitis was studied in 42 Japanese patients by Tagawa.[9] The only antigen significantly increased in this disease, which occurs more frequently in Japanese than in Caucasians, was a B-locus antigen Bw22J (Sa1), a Japanese variant of Bw22, which occurred with a 45.2% frequency in patients compared to 13.2% in 76 healthy controls (corrected $P < 0.02$), yielding a relative risk of 5.4. There was also an increased frequency of the D-locus antigen known as Wa, possibly associated with Sa1, which occurred with a frequency of 66.6% compared to 16.0% in 81 healthy controls, yielding a relative risk of 10.5.[10] However, the Bw22J association was unable to be confirmed in nine other patients, none of whom were Japanese.[11]

E. Cogan's Syndrome

This syndrome of nonsyphilitic interstitial keratitis and vestibuloauditory symptoms was associated with Bw17 in four of five patients reported.[12,13] However, a later study of 10 patients failed to confirm this.[14]

F. Behçet's Syndrome (see Dermatology [Chapter 7, Section V])

G. Diabetic Retinopathy (see Endocrinology [Chapter 7, Section VI])

H. Exophthalmos (see Endocrinology [Chapter 7, Section VI])

I. Retinoblastoma (see Malignancy [Chapter 7, Section IX])

J. Optic Neuritis (see Neurology and Psychiatry [Chapter 7, Section X])

REFERENCES

1. **Ryder, L.P. and Svejgaard, A.,** Associations Between HLA and Disease, Report from the HLA and Disease Registry of Copenhagen, HLA and Disease Registry, Copenhagen, 1976.
2. **Brewerton, D.A., Caffrey, M., Nicholls, A., Walters, D., and James, D.C.,** Acute anterior uveitis and HL-A27, *Lancet,* 2, 994, 1973.
3. **Brewerton, D.A. and James, D.C.O.,** The histocompatibility antigen (HL-A27) and disease, *Semin. Arthritis Rheum.,* 4, 191, 1975.
4. **Woodrow, J.C., Mapstone, R., Anderson, J., and Usher, N.,** HL-A27 and anterior uveitis, *Tissue Antigens,* 6, 116, 1975.
5. **Khan, M.A., Braun, W.E., and Kushner, I.,** Low frequency of HLA-B27 in American Blacks with ankylosing spondylitis, *Clin. Res.,* 24, 331A, 1976.
6. **Shin, D. and Becker, B.,** HLA-B12 and HLA-B7 in primary open-angle glaucoma, in *HLA and Disease,* INSERM, Paris, 1976, 289.
7. **Aviner, Z., Henley, W.L., Fotino, M., and Leopold, I.H.,** Histocompatibility (HL-A) antigens and primary open-angle glaucoma, *Tissue Antigens,* 7, 193, 1976.
8. **Becker, B. and Shin, D.,** HLA-B12 and HLA-B7: Important prognostic factors in ocular hypertension, in *HLA and Disease,* INSERM, Paris, 1976, 290.

9. **Tagawa, Y., Sugiura, S., Yakura, H., Wakisaka, A., Aizawa, M., and Itakura, K.,** Major histocompatibility antigens and Vogt-Koyanagi-Harada syndrome, in *HLA and Disease,* INSERM, Paris, 1976, 263.
10. **Yakura, H., Wakisaka, A., Aizawa, M., Itakura, K., Tagawa, Y., and Sugiura, S.,** HLA-D antigen of Japanese origin (LD-Wa) and its association with Vogt-Koyanagi-Harada syndrome, *Tissue Antigens,* 8, 35, 1976.
11. **Ohno, S., Char, D.H., Kimura, S.J., and O'Connor, G.R.,** HLA and Vogt-Koyanagi-Harada syndrome, *N. Engl. J. Med.,* 295, 788, 1976.
12. **Char, D.H., Cogan, D.G., and Sullivan, W.R., Jr.,** Immunologic study of nonsyphilitic interstitial keratitis with vestibuloauditory symptoms, *Am. J. Ophthalmol.,* 80, 491, 1975.
13. **Del Carpio, J., Espinoza, L.R., and Osterland, C.K.,** Cogan's syndrome and HLA-Bw17, *N. Engl. J. Med.,* 295, 1262, 1976.
14. **Kaiser-Kupfer, M. I., Del Valle, L. A., Mittal, K. K., and Haynes, B. F.,** HLA and Cogan's syndrome, *N. Engl. J. Med.,* 298, 1094, 1978.

XII. PULMONARY DISEASES

A. Silicosis

A study of 75 patients with silicosis (38 miners, 35 porcelain workers, and 2 quarrymen) compared their HLA antigen frequencies to those of 160 normal healthy individuals living in the same area of France, as well as to 46 other subjects who had been exposed for at least 20 years to the same risk of silicosis but had developed no clinical or radiological evidence of the disease.[1] The patients with silicosis had a decreased frequency of B7 to 6.7% compared to 20 and 21.3% among the normal and the other control group, respectively. There was a greater frequency of B8 in patients who had contracted tuberculosis (35.1%). However, neither of these deviations was significant.

B. Asbestosis

Among 134 asbestos workers with at least 20 years of exposure, there were 22 with radiographic and clinical evidence of asbestosis and 112 without signs of this disease.[2] Only 9.8% of those without asbestosis had B27 compared to 27.3% of those with asbestosis, yielding a relative risk of 3.4.[2]

Similar increases in B27 were also reported by Merchant, who studied 56 selected asbestos workers with definite or suspected asbestosis. B27 occurred in 17.9% of the patients compared to 5.2% of controls, a difference which was not significant when the P value was corrected for the number of antigens tested.[3] There was no significant association between B27 and the severity of asbestosis, since just 6 of 10 patients with B27 had moderate to severe disease compared to 13 of 46 without B27. Those with B27 tended to have a shorter exposure before developing asbestosis than did those without B27. An important fact was that none of the ten subjects with B27 had clinical or radiologic evidence of ankylosing spondylitis.

C. Farmer's Lung

A comparison of the frequency of HLA antigens in 20 patients with farmer's lung revealed that 40% of the farmer's lung patients had B8 compared to 8% of the controls.[4] However, because the control group numbered only 42 and the usual Caucasian frequency for B8 is approximately 20%, these data are tenuous.

D. Cryptogenic Fibrosing Alveolitis

Of 15 patients, 13 males and 2 females, 12 were found to have B12.[5] Although the 80% frequency far exceeded the 30% found in 616 controls, a subsequent study showed an increased frequency of B8 in females and those less than 50 years of age at the onset of their disease.[6]

E. Pneumoconiosis

Four patients with possible pneumoconiosis were among eight persons carrying the S allele of the proteinase inhibitor (Pi), the former having a strong association with A28 and A29 and the latter being linked to Gm allotypes. Of the eight unrelated people who had the S allele of Pi, 75% had the A28 antigen (compared to 6.5% of 276 controls) and 50% had A29 (compared to 2.5% of the controls).[7] The chances of finding A28 and/or A29 in the S allele group was 100% compared to 9% of the controls, a highly significant difference ($P < 10^{-8}$).[7]

F. Sarcoidosis

An early study by Kueppers showed that no significant antigen deviation was found in 132 American patients with sarcoidosis compared to 600 controls.[8] Fifty Swedish patients with sarcoidosis were found to have a 52% frequency of B7 (compared to 27% in 100 controls).[9] The phenotype A2-B7 was found in 28% (14/50) of sarcoid patients compared to 14% of controls. However, neither of these differences was significant. In the 132 patients studied by Kueppers, B7 was only slightly increased (30.4% compared to 26.2% of controls).[10]

A study of sarcoidosis and tuberculin sensitivity in that disease was conducted by Persson.[11] The B7 antigen again was found not to be increased in 80 cases of sarcoidosis. Yet, in the group with a negative tuberculin reaction following the appearance of the disease, there was a significant increase in B7 compared to controls (46.8% and 26.8%, respectively), whereas in patients with a positive tuberculin reaction, there was a complete absence of B7. Of the ten patients who were symptomatic, seven had the B7 antigen. It was noted that the Swedish patients studied in 1973 by Hedfors were all tuberculin-negative, and in the study by Kueppers, delayed hypersensitivity was absent or diminished.[11] Thus, these authors concluded that the HLA system did not seem to alter the likelihood of developing sarcoidosis, but once the condition developed, B7-positive individuals were more likely to have depressed cellular immunity to tuberculin and to demonstrate clinical symptoms.[11]

Another study of sarcoidosis in 117 Japanese patients with the disease showed only a suggestive increase of B13.[12]

G. Hypersensitivity Pneumonitis (see Allergy [Chapter 7, Section I])

H. Asthma (see Allergy [Chapter 7, Section I])

I. Bronchogenic Carcinoma (see Malignancy [Chapter 7, Section IX])

REFERENCES

1. **Gualde, N., deLeobardy, J., Serizay, B., and Malinvaud, G.,** HLA and silicosis, in *HLA and Disease,* INSERM, Paris, 1976, 248.
2. **Matej, H. and Lange, A.,** HLA antigens in asbestosis, in *HLA and Disease,* INSERM, Paris, 1976, 256.
3. **Merchant, J.A., Klouda, P.T., Soutar, C.A., Parkes, W.R., Lawler, S.D., and Turner-Warwick, M.,** The HL-A system in asbestos workers, *Br. Med. J.,* 1, 189, 1975.

4. **Flaherty, D.K., Iha, T., Chmelik, F., Dickie, H., and Reed, C.E.**, HL-A8 in farmer's lung, *Lancet*, 2, 507, 1975.
5. **Evans, C.C. and Evans, J.M.**, HL-A in farmer's lung, *Lancet*, 2, 975, 1975.
6. **Turton, C. W., Morris, L. M., Lawler, S. D., and Warwick, M. T.**, HLA in cryptogenic fibrosing alveolitis, *Lancet*, 1, 507, 1978.
7. **McIntyre, J., Ainsworth, S., Allen, R., Loadholt, C., and Reigart, J.**, Increased frequency of HLA-A28 and A29 in alpha-1-antitrypsin phenotypes having the S allele, in *HLA and Disease*, INSERM, Paris, 1976, 297.
8. **Kueppers, F., Brackertz, D., and Mueller-Eckhardt, Ch.**, HL-A antigens in sarcoidosis and rheumatoid arthritis, *Lancet*, 2, 1425, 1972.
9. **Hedfors, E. and Möller, E.**, HL-A antigens in sarcoidosis, *Tissue Antigens*, 3, 95, 1973.
10. **Kueppers, F., Mueller-Eckhardt, Ch., Heinrich, D., Schwab, B., and Brackertz, D.**, HL-A antigens of patients with sarcoidosis, *Tissue Antigens*, 4, 56, 1974.
11. **Persson, I., Ryder, L.P., Staub Nielsen, L., and Svejgaard, A.**, The HL-A7 histocompatibility antigen in sarcoidosis in relation to tuberculin sensitivity, *Tissue Antigens*, 6, 50, 1975.
12. **Hoshino, K., Inouye, H., Unokuchi, T., Ito, M., Tamaoki, N., and Tsuji, K.**, HLA and diseases in Japanese, in *HLA and Disease*, INSERM, Paris, 1976, 249.

XIII. RENAL DISEASES

A. Glomerulonephritis

A study in 1969 by Patel, Mickey, and Terasaki on the association of A2 with chronic glomerulonephritis was one of the earliest assays of an association between an HLA antigen and disease.[1] These authors reported a 52% incidence of A2 in patients with chronic glomerulonephritis of a wide variety of subtypes compared to 42.3% of controls. This tempting association was a very far-sighted venture but one that has never been confirmed and in fact, has been contradicted recently.[2] This early report provided several illustrative points in disease association studies which only became clear with the passage of time. First of all, the antigen that the patients were suspected of having with a higher frequency, A2, happens to be the most common antigen in the Caucasian population, a fact which makes truly significant increases difficult to establish. Secondly, the clinical entity which they were investigating, glomerulonephritis, is composed of a wide variety of subtypes and perhaps in many cases, actually represented misdiagnoses based on the evaluation of end-stage kidneys or only suggestive clinical histories. Third, the accuracy of tissue typing, although it was certainly as great in the hands of these authors as those anywhere in the world, in terms of today's typing was very incomplete. Therefore, numerous other A- and B-series antigens could well have been included in reagents that were being used for the identification of A2. Fourth, the P values were not corrected for the number of antigens investigated, although it may have been that the A2 antigen was actually the only one examined.

Despite these early equivocal results, very recent studies of specific subtypes of glomerulonephritis have been more revealing.[3-5] In a study of 121 patients subgrouped as mesangiocapillary glomerulonephritis, membranous glomerulonephritis, diffuse proliferative glomerulonephritis, mesangial proliferative glomerulonephritis, focal and segmental glomerulonephritis, and focal and segmental hyalinosis, MacDonald reported an association of membranous glomerulonephritis with Bw40, mesangial proliferative glomerulonephritis with Bw35, and focal and segmental glomerulonephritis with B12, each with a corrected $P < 0.04$.[3] In 246 patients having membranoproliferative glomerulonephritis, IgA mesangial deposits (Buerger's disease), focal and segmental glomerulosclerosis, or Alport's syndrome, Bw35 was found in 47% of those with Buerger's disease compared to 19% of the normal controls (corrected $P < 0.03$).[4]

Another extensive study of 105 patients with the nephrotic syndrome, chronic glomerulonephritis, Henoch-Schonlein nephritis, or other forms of glomerulonephritis showed an increased frequency of B5 and B15 in chronic glomerulonephritis and of Bw35, B18, and B27 in Henoch-Schonlein purpura, none of these being significant, however. The common feature of the Bw35-associated glomerulopathies (mesangial-proliferative, Buerger's disease, and Henoch-Schonlein purpura) appears to be IgA deposition in the glomeruli. Possibly related is the finding that an A11-Bw35 haplotype was found in 5 of 19 cases of the complex form of necrotizing venulitis, but not in any of 12 instances of the simple form.[6]

It is interesting in these various forms of glomerulonephritis and in necrotizing venulitis, which are possibly of viral etiology, that Bw35 or antigens cross reacting with it, such as B15, B5, and B18, generally have been found to be increased. The increased frequency of this cross-reacting group is similar to the findings in Hodgkin's disease, also of probable viral causation. Balkan nephropathy has also been reported to have an increased frequency of Bw35.[7] It has been noted in a negative fashion that no individual with true streptococcal infections had Bw35.[5]

A report of a family with C2 deficiency and renal disease in all four children showed that each was a homozygote for the A10-B18 haplotype.[8] The frequency of the A10-B18 haplotype, B18, and B40 were all increased in renal allograft recipients whose original disease was immune complex nephritis,[8] a finding supported by the fact that C2 deficient individuals have the A10-B18 haplotype (see Complement Deficiencies [Chapter 7, Section III]). The linkage of HLA with several complement components and properdin factor B, which are involved in immunologic renal injury, makes the major histocompatibility complex with or without HLA relevant to renal disease.

B. Lupus Nephritis (see Connective Tissue Diseases [Chapter 7, Section IV])

As noted under systemic lupus erythematosus (SLE), 66% of 17 patients with nephropathy due to SLE had B15 and Bw24, which were 12 and 4 times, respectively, more frequent than in the control population.[9] "Extra" antigens, usually in the B series, were also found in the acute phase of the disease. The latter finding suggested the appearance of B-cell antigens detected with antisera having more than just HLA reactivity — an hypothesis proposed by Terasaki for his own similar results in immunodeficiency diseases (see Immunodeficiency Diseases [Chapter 7, Section III]).[10]

C. Polycystic Kidney Disease

A preliminary report of 75 patients by Dausset showed an increase in B5 to 32% from 13% in controls (corrected $P < 0.01$).[11] Another study from France by Noel also demonstrated an increased frequency of B5 in an unspecified number of such patients (24% compared to 13% of controls, corrected $P < 0.05$).[4]

D. Steroid-responsive Nephrotic Syndrome

In Thomson's study of 71 Caucasian children who had steroid-responsive nephrotic syndrome (SRNS), a significant increase in B12 was found in the nephrotic children to 54% compared to 15% in 39 healthy adults and 24% in 1036 blood donors (corrected $P < 0.02$).[12] A complementing decrease of B7 to 7% in those with the nephrotic syndrome was significant when compared to the 26% frequency in blood donors (corrected $P < 0.02$), although it was not significantly different from the 23% frequency in healthy adults. Interestingly, atopic symptoms occurred in 13 of 35 nephrotic children with B12 (37%) in contrast to only 1 of 30 nephrotic children without B12 (3%; $P < 0.01$). The estimated risk for SRNS was 9.5 for children with atopy and B12 when compared to all others and 13 when compared to those without atopy or B12. Relapses after cyclophosphamide were also more frequent in those with B12 (12/15) than in those without (5/14). The A1-B8 phenotype was significantly increased in the nonatopic SRNS children.

E. Vesico-ureteric Reflux

Of 43 patients, 17 had A3, which was a significant increase (corrected $P < 0.003$).[13]

F. Analgesic Abuse Nephropathy

A study of 34 patients with analgesic abuse showed an increase in A3 (corrected $P < 0.002$).[13]

G. Familial Renal Cell Carcinoma

In a study of four unrelated families with renal cell carcinoma in at least two members, all four of the propositi and seven of the eight patients tested had B17.[14] This disease, which has an increased frequency in males and a late onset in life, lost its association with the B17 antigen when occurring in a random population. There was a remarkable parallel between the inheritance pattern of the disease in one of these four families and in a family with ankylosing spondylitis described by van der Linden that supported at least two genetic factors being involved.[15] In the family with renal cell carcinoma, the mother, who had B17, was free of disease whereas the father, with renal cell carcinoma, lacked B17 but presumably was homozygous for a recessive gene predisposing to renal cell carcinoma. One of the sons with B17 and presumably with one of the recessive genes which his father possessed also had renal cell carcinoma. It is the presumed non-HLA associated disease susceptibility gene that has escaped an identifying mark except for the occurrence of the disease.

H. Unspecified Renal Disease

A study from Australia of 306 patients with a variety of renal diseases revealed that A1 and A3 occurred with increased frequency (corrected $P < 0.02$) when compared to a normal population.[16] It was also interesting to note that the antigens typically found in linkage disequilibrium with these two A-series antigens, namely B8 and B7, were not found with increased frequency, thereby suggesting a greater importance of the A-series antigens A1 and A3 for renal disease.

I. Immune Complex Nephritis (see Complement Deficiency Diseases [Chapter 7, Section III])

J. Prospects of Further HLA Association and Renal Disease

From a theoretical point of view, the immune-complex types of renal disease may hold great promise for association with antigens in the major histocompatibility complex. Experimental work by Dixon showed that the type of renal disease that developed in animals in response to chronic antigen administration was determined by the magnitude of the antibody response in the immunized animal.[17] Those with no antibody response were essentially free of disease, those with an antibody response in the equivalence zone or slight antigen excess developed chronic glomerulonephritis, and those with a high antibody response developed an acute glomerulonephritis with only a small number developing chronic disease.[17] The feature which links this phenomenon to a future role for histocompatibility testing lies in studies which have very clearly pointed out that the magnitude of an antibody response can be determined by immune response antigens, referred to earlier as Ir and Ia antigens, located in or near the major histocompatibility complex in animals.[18] Consequently, if certain types of glomerulonephritis are determined by the level of antibody response that an individual makes in response to a certain antigen and if the responsiveness to relevant classes of antigens can be identified by typing for Ir or Ia antigens located in the major histocompatibility complex, then it seems promising that histocompatibility testing in its broad sense will be able to identify those individuals with a high risk of developing this type of renal disease. The increased responsiveness of those with B5 to streptococcal antigen offered preliminary experimental support for such an association being found.[19]

The mapping of loci for several complement components as well as properdin factor B near the HLA loci on the C6 chromosome as described earlier further enhances the prospects for such an association.

The role that HLA antigens may have in virus-induced diseases, among which are some renal diseases, is discussed in Possible Mechanisms of HLA and Disease Associations (Chapter 3).

REFERENCES

1. **Patel, R., Mickey, M.R., and Terasaki, P.I.,** Leukocyte antigens and disease: I. Association of HL-A2 and chronic glomerulonephritis, *Br. Med. J.,* 2, 424, 1969.
2. **Jensen, H., Ryder, L.P., Nielsen, L.S., Clausen, E., Jorgensen, F., and Jorgensen, H.E.,** HLA antigens and glomerulonephritis, *Tissue Antigens,* 6, 368, 1975.
3. **MacDonald, I.M., Dumble, L.J., and Kincaid-Smith, P.,** HLA and glomerulonephritis, in *HLA and Disease,* INSERM, 1976, 203.
4. **Noel, L.H., Descamps, B., Jungers, P., Bach, J.F., Busson, M., Guillet, J., and Hors, J.,** HLA serotyping in 5 well defined kidney diseases, in *HLA and Disease,* INSERM, 1976, 206.
5. **Nyulassy, S., Buc, M., Sasinka, M., Pavlovic, M., Hirschova, V., Kaiserova, M., and Stefancvic, J.,** HLA system in glomerulonephritis, in *HLA and Disease,* INSERM, Paris, 1976, 207.
6. **Carpenter, C.B.,** personal communication, 1976.
7. **Mowbray, J.F.,** personal communication, 1976.
8. **Mowbray, J.F.,** Association of heterozygous C2 deficiency with both disease and HLA, in *HLA and Disease,* INSERM, Paris, 1976, 204.
9. **Stefanova, G.,** Relationship between HLA and other immunological tests in nephropathy due to SLE, in *HLA and Disease,* INSERM, Paris, 1976, 211.
10. **Terasaki, P.I.,** personal communication, 1976.
11. **Dausset, J. and Hors, J.,** Some contributions of the HL-A complex to the genetics of human disease, *Transplant. Rev.,* 22, 44, 1975.
12. **Thomson, P.D., Barratt, T.M., Stokes, C.R., Turner, M.W., and Soothill, J.F.,** HLA antigens and atopic features in steroid-responsive nephrotic syndrome of childhood, *Lancet,* 2, 765, 1976.
13. **MacDonald, I.M., Dumble, L.J., and Kincaid-Smith, P.,** HLA A3, vesico-ureteric reflux and analgesic abuse, in *HLA and Disease,* INSERM, Paris, 1976, 255.
14. **Braun, W.E., Strimlan, C.V., Negron, A.G., Straffon, R.A., Zachary, A.A., Bartee, S.L., and Grecek, D.R.,** The association of W17 with familial renal cell carcinoma, *Tissue Antigens,* 6, 101, 1975.
15. **van der Linden, J.M., Keuning, J.J., Wursman, J.H., Cats, A., and van Rood, J.J.,** HL-A27 and ankylosing spondylitis, *Lancet,* 1, 520, 1975.
16. **Dumble, L.J., MacDonald, I.M., and Kincaid-Smith, P.,** HLA antigens and renal disease, in *HLA and Disease,* INSERM, Paris, 1976, 246.
17. **Dixon, F.J., Feldman, J.D., and Vazquez, J.J.,** Experimental glomerulonephritis: the pathogenesis of a laboratory model resembling the spectrum of human glomerulonephritis, *J. Exp. Med.,* 113, 899, 1961.
18. **Sheffler, D.C. and David, C.S.,** The H-2 major histocompatibility complex and the immune response region: genetic variation, function, and organization, *Adv. Immunol.,* 20, 125, 1975.
19. **Greenberg, L.J., Gray, E.D., and Yunis, E.J.,** Association of HL-A5 and immune responsiveness in vitro to streptococcal antigens, *J. Exp. Med.,* 141, 935, 1975.

XIV. RHEUMATOLOGY (INCLUDING MUSCULOSKELETAL DISEASES)

A. Ankylosing Spondylitis

The primary antigen shown to occur with increased frequency in this disease is B27 (combined uncorrected $P < 1.0 \times 10^{-10}$) with a relative risk (RR) of 87.78 and 95% confidence interval of 68.20 to 112.99.[1] The major studies of Caucasians, American Blacks, Indians, Japanese, and Jews with this disease are shown in Tables 21 and 22.

It has been in this area that the most dramatic association has been found between an HLA antigen and a particular disease. Nearly simultaneous studies by Schlosstein and Brewerton[2,3] in 1973 demonstrated approximately an 11- to 24-fold increase (8 to 88% and 4 to 96%, respectively) in the frequency of the HLA antigen B27 in patients with ankylosing spondylitis (AS). This remarkable association, which is the strongest found thus far between any HLA antigen and a disease, confers a relative risk of approximately 88 times for a Caucasian.[1] Although there is some variability among the studies that may be due to methods of patient selection and clinical criteria for the disease, this extremely strong association has been repeatedly confirmed in Caucasians (Table 21).

However, the strength of the association may diminish in other populations. For example, combined studies of American Blacks with ankylosing spondylitis showed that only 51 of 76 patients (67.1%) [2,4-11] had the B27 antigen with a relative risk in the

TABLE 21

Ankylosing Spondylitis in Caucasians and Jews

HLA Ag	Controls with Ag (%)	Patients with Ag (%)	Patients with Ag (n)	Patients total (n)	Relative risk	Ref.
Caucasians						
B27	4.0	96.0	72	75	429	3
	8.0	87.5	35	40	74	2
	3.9	81.1	43	53	93	36
	5.0	76.9	20	26	58	37
	9.2	90.0	45	50	81	38
	12.4	100.0	35	35	485	39
	8.9	82.4	14	17	42	40
	9.8	94.0	63	67	127	33
	7.8	91.5	43	47	111	41
	12.0	84.1	106	26	37	24
	8.7	91.4	32	35	95	42
	9.4	94.0	47	50	129	43
	9.1	96.3	26	27	175	11
	5.5	85.0	68	80	97	44
	6.7	96.8	60	62	335	45
	8.4	94.0	47	50	179	4
Jews						
B27						
Ashkenazi	—	90.0	18	20	281	17
Non-Ashkenazi	—	75.0	12	16	91	17

TABLE 22

Ankylosing Spondylitis in Non-Caucasians

HLA Ag	Controls with Ag (%)	Patients with Ag (%)	Patients with Ag (n)	Patients total (n)	Relative risk	Ref.
American Blacks						
B27	4.0	80.0	8	10	18.3	2
	4.0	42.9	3	7	97.6	7
	—	100.0	9	9	—	5
	1.7	47.8	11	23	54	4[a]
	4.0	47.8	11	23	23	4[a]
	—	60.0	9	15	—	8
	—	88.9	8	9	—	6
Indians (Haida, Bella Coola, and Pima)						
B27						
Haida	50.5	100	17	17	34	13
Bella Coola	25.6	100	3	3	20	13
Pima	18.0	26.3	10	38	—	14
Japanese						
B27						
Japanese (Singapore)	0.0	66.7	18	27	306	15

[a] The different results are based on two control groups that are not significantly different.

largest series of between 23 and 54[4] according to the control group used.[4] It should be noted that when one collects the cases of Blacks with ankylosing spondylitis, there is a tendency for the results to be skewed toward an association with the antigen because of the inclusion of single cases reported primarily because of their positive association.[9-11] In African Blacks, the frequency of B27 was close to zero and the disease of ankylosing spondylitis was rare; four of six cases showed the B27 antigen.[12]

In the Bella Coola and the Haida Indians of British Columbia, the B27 frequency rose from 25.6 and 50.5% in respective controls to 100% in a total of 20 patients with AS. Because of the unusually high frequency of B27 in controls, however, the relative risk increased to only 20.2 and 34.4, respectively.[13] In the Pima Indian population, in which the frequency of B27 in controls was 18%, only 26.3% of the 38 with AS had B27.[14]

In the Japanese population, despite the virtual absence of B27, those with AS usually had B27 (18/27 or 66.7%).[15] Of 16 Singapore Chinese with AS, 15 had B27.[16] Whether this represents a Caucasian or an Indian contribution of B27 to the Japanese and Chinese is uncertain.

In the Israeli population, the frequency of B27 in AS was less in non-Ashkenazi (75% of 16 cases) than in Ashkenazi Jews (90% of 20 cases).[17] Approximately 82% of 38 Sardinian patients with AS had B27 compared to 5.3% in 494 controls.[18]

Aside from the fact that the diagnostic value of B27 for AS is less in non-Caucasians (see Clinical Use of HLA Typing [Chapter 5]), the diminished association with AS in

American Blacks, African Blacks, Japanese, and the Pima Indians also indicates that the gene for B27 is not the sole susceptibility site or factor for AS (see also Animal Studies as the Basis of Disease Association and Immune Responsiveness [Chapter 2]).

Other evidence from family studies suggests at least one other gene, and possibly more, as being necessary for the development of AS. An early study by Espinoza that showed about a 49% occurrence of B27 but only 1 case of AS among 37 relatives of patients with AS pointed out that B27 was not synonymous with AS and that a second factor was necessary.[9] A family from the Netherlands was described in which the mother had B27 but did not have AS, the father did not have B27 but had AS and was the offspring of a cousin marriage and therefore could have been homozygous for a recessive gene, and the children who developed AS all created a pattern compatible with a recessive gene being inherited from each parent, one in association with B27 and the other one not.[19] Another family had two HLA-identical siblings with B27, one with and one without AS, as well as another sibling without B27 but with AS.[20] This pattern could have resulted either from a polygenic effect or recombination.

Two sets of identical twins discordant for B27 positivity and AS demonstrated that whether a polygenic effect is essential for the development or nondevelopment of AS, an environmental factor is also critical.[21] A study of 64 members in three generations of another family with a high incidence of AS had the following results: All AS patients had B27 but not all B27-positive members had AS, thus confirming the strong B27-AS association but showing that they are not synonymous and may require another factor.[22] Only the 1 homozygous B27 daughter, but not the 13 heterozygous females developed AS, whereas all 3 homozygous and 3 of 16 heterozygous males had AS, thus confirming the male predominance in AS and suggesting added potency of B27 homozygosity in causing the disease in females.

Further data concerning the influence of sex on the severity of AS are brought out in a study by Jeannet, who found B27 in 87.1% of females with AS in whom nearly half had no radiographic evidence of disease in the dorsolumbar spine.[23] Why females appear to have a milder form of AS is uncertain, but possible reasons for the influence of sex on disease occurrence and HLA antigens were discussed earlier (see Possible Mechanisms of HLA and Disease Associations: Sex Prevalence [Chapter 3]). Concerning the frequency of B27 in Black females, conflicting data that is probably the result of very small sample size were presented by Swezey,[7] who reported only three of seven to be positive, and Levitin,[6] who found eight of nine to have B27.

The evidence concerning the severity of AS in B27-positive patients is also conflicting. Möller noted that the frequency of B27 jumped from 70% in 60 patients with sacroiliitis to 100% in the 66 with pelvospondylitis.[24] However, the failure to exclude other diseases associated with sacroiliitis which are known to diminish the B27 association may have been responsible for a lower B27 frequency in that condition. Good's study in 15 Blacks indicated that the disease was more severe in B27-positive patients,[8] whereas Khan's study of a larger and more representative group of Blacks did not find any difference in disease severity in Blacks, though iritis was more common in B27-positive patients, particularly Blacks.[4] Svejgaard also reported no additive adverse effect of B27 in AS patients who were homozygous for B27,[25] although the occurrence in one family of AS only in the B27 homozygous female and in none of 13 heterozygous females suggests that its effect may be sex-limited.[22]

The finding that a disease accompanied by AS, idiopathic aortic insufficiency, failed to show any association with B27 was somewhat of a surprise.[26] However, this result fits with the general finding that the strength of the B27-AS association is always diminished when another disease accompanies AS.

The most provocative area is the exploration of the significance of B27 in a healthy population. A study of 78 "healthy" blood donors who possessed B27 revealed that

14 had AS (18%).[27] The high frequency of AS in this study in both male and female B27-positive healthy persons is in contrast to the generally accepted frequency of 0.2% in females and 2% in males. Not a single case of AS was found in 126 B27-negative controls matched for age, race, and sex. Similar striking results were found in another smaller study of 24 apparently healthy men, 6 of whom had either definite or strongly suggestive evidence of AS.[28] Whether the 33% percentage of AS in first degree relatives of AS patients is significantly different from the 20% reported in "healthy" B27-positive blood donors is unknown at present.[29]

Another interesting aspect of the association of B27 with ankylosing spondylitis in Caucasians has been the finding by Terasaki and Mickey that an antigen of the A series (A2) occurs much more frequently with B27 than would be expected, so that the A2-B27 haplotype appears to be even more crucial to the occurrence of ankylosing spondylitis than does the B27 antigen alone.[30] Supporting this were the findings in a series of 27 AS patients in Yugoslavia that the A2-B27 haplotype occurred in 33.5% of patients and only 3.2% of controls and that 86% of the A2 genes were coupled to B27 in the patients compared to 13% of controls.[31]

The association of AS with C- and D-series antigens was studied by Dausset, who found that 53% of 53 cases of AS had Cw2.[32] However, none of the five unrelated patients tested shared D antigens as reflected in high stimulation in MLCs among these five patients. In Bruyere's study, MLCs among four HLA-identical siblings, only one with AS, were all nonstimulatory, thus showing that the D-locus antigens measured by MLC were the same whether AS was present or absent.[22] Möller[24] and Sachs[33] have also failed to find a common D-series antigen with AS even when B27 homozygosity existed.[24]

The B27 antigen association of AS has been taken advantage of in order to search for a possible etiologic agent for AS with the working hypothesis that AS is an aberrant response to some infectious agent. A rabbit antiserum to B27-positive lymphocytes reacted only against Klebsiella aerogenes when tested by immunodiffusion and by hemagglutination against 27 different strains of microorganisms.[34] Conversely, antiserum raised against Klebsiella aerogenes had precipitating antibodies against B27-positive lymphocytes from patients with AS. This bidirectional cross reactivity suggested that Klebsiella may mimic the molecular configuration of the B27 antigen and trigger an aberrant host response in some unknown way that can lead to AS when other conditions are also fulfilled.

Another recent related finding has been the depressed T-cell activity that exists not only in B27-positive patients with AS, but also in B27-positive relatives without AS, as well as in AS patients who are B27-negative and their siblings and offspring who are B27-negative and have no AS.[35]

B. Reiter's Disease

The primary antigen shown to occur with increased frequency in this disease is B27 (combined uncorrected $P < 1.0 \times 10^{-10}$) with a relative risk of 35.89 and a 95% confidence interval of 25.77 to 49.98.[1] The major studies in Caucasians and American Blacks with this disease are shown in Table 23.

Reiter's disease may be defined as a form of peripheral arthritis associated with nongonococcal urethritis and conjunctivitis predominantly occurring in young males and sometimes associated with chronic sacroiliitis and spondylitis as well as aortic insufficiency. The initial HLA study of Reiter's disease by Brewerton described 33 patients with Reiter's disease, 25 of whom had the B27 antigen.[46] Only 3 of 33 patients with nonspecific urethritis (9%) and 2 of 33 controls (6%) had B27. In Morris' study of 24 patients with Reiter's syndrome, 23 of the 24 patients had B27 (96%) as compared to just 8% in 1863 Caucasian controls.[47] Although numerous subsequent studies

TABLE 23

Reiter's Syndrome in Caucasians and American Blacks

HLA Ag	Controls with Ag (%)	Patients with Ag (%)	Patients with Ag (n)	Patients total (n)	Relative risk	Ref.
			Caucasians			
B27						
	6.1	75.8	25	33	48	46
	8.0	95.8	23	24	180	47
	3.9	80.4	37	46	89	36
	7.7	68.0	34	50	24	51
	4.0	63.3	19	30	43	48
	14.1	91.8	56	61	62	52
	6.1	74.6	47	63	36	49
	8.2	64.6	31	48	20	53
	9.1	73.9	17	23	27	11
	7.8	100.0	6	6	150	41
			American Blacks			
B27						
	—	50.0	1	2	—	48
	—	14.3	1	7	—	8
	1.7	33.3	2	6	—	4[a]
	4.0	33.3	2	6	—	4[a]

[a] The different results are based on two control groups that are not significantly different.

have confirmed the increased frequency of B27 with Reiter's syndrome, no comparably sized study has shown the magnitude of B27 deviation that Morris' early study demonstrated (Table 21). An explanation for the lower frequency (63%) of the B27 antigen offered by McClusky in a study of 30 Reiter's patients is that the group studied by Morris had suffered from the disease for at least one full year and had established a chronic course.[48] Therefore, he suggested that B27 may occur in approximately two thirds of patients with Reiter's syndrome in the acute stage but in greater than 90% of those with a chronic and relapsing course, as in Morris' study.

In a later study, Brewerton and James described a total of 63 patients with Reiter's disease, 47 of whom had the B27 antigen in comparison to 3 of 33 patients with non-specific urethritis.[49] Their study also provided some interesting subdivisions of Reiter's disease and examined the occurrence of B27 within those subgroups. For example, in patients with Reiter's disease who had peripheral arthritis but no sacroiliitis, only 25 of 40 patients had B27, whereas sacroiliitis patients without spondylitis showed the B27 antigen in 13 of 14 cases, and all 9 of the patients with sacroiliitis and spondylitis showed the B27 antigen. This interesting progression of B27 positivity as the arthritis centrally progressed, going from the periphery to the sacroiliac joints to the spine itself, was also noted in the arthritis associated with diseases such as ulcerative colitis, Crohn's disease, and psoriasis, which will be described subsequently. Brewerton has observed that the typical frequency of 65% for B27 in Reiter's disease increases to more than 90% when balanitis, keratoderma blennorrhagica, or uveitis occurs with it.[35]

Of 43 patients with Reiter's syndrome, 57% had the C-series antigen Cw2, a reflection of the linkage disequilibrium between the B27 antigen and Cw2.[32]

The usefulness of the B27 antigen in the differential diagnosis of arthritis in the male in whom urethral symptoms are also present is demonstrated further by the fact that there was not a single instance of the B27 antigen in 12 patients with gonococcal arthritis, nine of whom were proved by culture, although it might be expected to occur in approximately 8% of a Caucasian and 3% of an American Black control population.[47] Gonococcal arthritis is discussed later in this section.

A different clinical presentation of Reiter's features has been labeled the "incomplete Reiter's syndrome."[50] In this particular clinical condition, the patient may have only peripheral arthritis without conjunctivitis or urethritis. The arthritis is typically oligoarticular, asymmetric, predominantly in the lower extremity, and frequently associated with heel pain, "sausage digits," and periostitis in other sites. The course was typically a chronic one. Studies of the HLA antigens revealed that 12 of the 13 patients falling into this clinical group had the B27 antigen.[50] Under these circumstances, the finding of the B27 antigen was clearly a diagnostic aid in establishing the fact that these patients had an incomplete form of Reiter's disease rather than some other form of arthropathy.

There are only a few brief reports of Reiter's syndrome in American Blacks, but these small groups again suggest that the frequency of B27 will be less in American Blacks (26.7%, 4/15) with Reiter's syndrome than in Caucasians (63 to 96%).[4,8,48]

The circumstances under which Reiter's syndrome has occurred have suggested a venereal mode of transmission, particularly in the U.S. and England, but no infectious agent has been firmly established as the causative factor. The development of Reiter's disease, particularly in Scandinavia and Europe after infectious dysentery caused by Shigella, is described under Postinfectious Arthropathies (Chapter 7, Section XIV.C). This particular form of postinfectious arthritis with Reiter's syndrome subsequently flows immediately into the next group of postinfectious arthropathies, also called reactive arthropathies.

C. Postinfectious Arthropathies (Reactive Arthropathies)

The primary antigen shown to occur with increased frequency in these diseases is B27. The major studies of these diseases are shown in Table 24.

TABLE 24

Postinfectious (Reactive) Arthropathies

HLA Ag	Controls with Ag (%)	Patients with Ag (%)	Patients with Ag (n)	Patients total (n)	Relative risk	Ref.
		Post-Shigella Arthropathy				
B27	—	78.0	39	50	—	56
	—	80.0	4	5	—	57
		Post-Salmonella Arthropathy				
B27	8.1	60.0	3	5	16	61
	10.1	69.2	9	13	19	59
	—	93.3	14	15	—	58
		Post-Yersinia Arthropathy				
B27	14.1	87.8	43	49	40	52
	8.6	57.9	11	19	14	1
	14.1	65.2	30	46	—	60

During World War II, there was an epidemic of Shigella flexner dysentery in Finland. Within a few weeks, 0.2% of the patients who had had dysentery developed Reiter's syndrome. Of 344 patients, 70% had urethritis, arthritis, and conjunctivitis.[54] When 100 patients were reevaluated 20 years later, 32 had spondylitis, 18 chronic peripheral arthritis, and 7 iritis.[55] Preliminary studies of the HLA antigens in these 100 patients showed that 39 of 50 patients (78%) had B27 that was more frequent in the more severe form of the disease, ranging from 70% in the mild form (19 of 27) to 94% when the sacroiliac joints or spine was involved (15 of 16).[56]

Another long-term follow-up study has been performed on individuals infected with Shigella in 1962 while on a naval ship. Of the ten original cases, five were able to be traced and four of the five, all of whom had a chronic course, were B27-positive. The single patient who was B27-negative had the mildest course and has been symptom-free for the past 2 years. Thus, in postinfectious Reiter's disease, just as in typical Reiter's disease, the presence of the B27 antigen appears to be associated with a more chronic and aggressive course.[57]

Following salmonella infections, a peripheral arthritis may develop in approximately 2% of patients. Preliminary results from a Finnish study revealed that 14 of 15 patients with post-salmonella arthritis had the B27 antigen.[58] No information is currently available on the late manifestations of the disease in terms of sacroiliitis or spondylitis. In a Swedish outbreak of 330 cases of Salmonellosis due to *Salmonella typhimurium*, 9 of 13 cases with arthritis had B27, or 69.2% compared to 7.7% of controls. No evidence of septicemia, purulent arthritis, or cross-reacting antibodies to *Yersinia enterocolitica* was found.[59]

Another infectious agent that has a close association with the B27 antigen is *Yersinia enterocolitica,* which causes an arthritis similar to postdysenteric Reiter's disease. However, in this type of arthritis, the sex distribution is equal and the skin, mucous membranes, and eyes are spared. A follow-up evaluation of 49 patients from Finland who had their attack between 1968 and 1970 showed that 43 had B27 (87.7%).[52] In a recent study of Yersinia arthritis from Finland, 30 of 46 cases (65%) had B27.[60]

Although the presence of the B27 antigen offers both diagnostic and prognostic information in these diseases, it is necessary to recall that there is approximately a 35% false-negative result in Reiter's disease and reactive arthritis.

D. The Arthropathies of Inflammatory Bowel Diseases (IBD)

In the arthropathies associated with inflammatory bowel disease, the first type of arthropathy, a peripheral one, has its onset with or generally after the development of the colitic symptoms. It primarily affects the lower extremities asymmetrically and shows a high degree of correlation with the activity of the bowel disease. The second type of arthropathy, a spondylitis, sometimes occurs before the onset of overt bowel disease and is usually unrelated to the activity, severity, or chronicity of the bowel disease. Furthermore, the activity of the spondylitis is not influenced by medical or surgical treatment of the bowel disease.[62]

It has been estimated that patients with ulcerative colitis had ankylosing spondylitis up to 30 times as often as in the general population.[49] In relatives of patients with ulcerative colitis who themselves have no rheumatic disease, the occurrence of an increased incidence of sacroiliitis and spondylitis has been attributed to the common factor of B27.[63]

Ulcerative colitis and Crohn's disease by themselves have no strong deviation in HLA antigen frequency (see Gastroenterology [Chapter 7, Section II]). As arthritis centripetally progressed, the B27 frequency increased.[49] In ulcerative colitis, only one of eight patients was positive for B27 when peripheral arthritis alone was present. Only one of three patients was positive for B27 when sacroiliitis was present without spondylitis. However, 13 of 18 patients with ulcerative colitis who had sacroiliitis and spondylitis were positive for B27. Similarly, in Crohn's disease, only one of four patients with peripheral arthritis alone and one of three patients with sacroiliitis but without spondylitis had the B27 antigen. However, four of five patients who had Crohn's disease with sacroiliitis and spondylitis had the B27 antigen. The increasing frequency of the B27 antigen as one progresses toward the spine is similar to that described earlier in Reiter's disease.[49]

In another study of patients with inflammatory bowel disease, only those who had spondylitis had the B27 antigen (6 of 8 cases).[64] Of the eight patients with spondylitis, four (50%), all B27-positive, had iritis compared to just one B27-negative patient without spondylitis (4%).

A study from Belgium revealed that in 21 patients with regional enteritis, 11 of whom had classical AS, 4 possible AS, and 6 asymptomatic sacroiliitis, only 3 (14%) had B27, all of whom were in the classical AS group.[65] This type of finding in which a normally high-frequency antigen in ankylosing spondylitis, namely, B27, shows a substantial decrease in frequency when one adds a disease such as ulcerative colitis or Crohn's disease suggests that something in the gastrointestinal tract lesion might alter a possible infectious etiologic agent for AS in such a way as to modify its effectiveness in inducing AS. Another possibility is that an undiscovered HLA, Ir, or Ia antigen associated with the inflammatory bowel disease in some way interferes with the susceptibility created by the B27 antigen.

E. Psoriatic Arthritis (see Dermatology [Chapter 7, Section V])

The first antigens shown to occur with increased frequency in this disease are B27, B13, Bw17, and Bw38. In the central form of this arthropathy, these antigens have combined uncorrected P values of $< 1.0 \times 10^{-10}$, 7.6×10^{-8}, 4.6×10^{-3}, and 5.6×10^{-7}, and relative risks of 8.58, 4.79, 2.49, and 9.09, respectively. In the peripheral form, these antigens have combined uncorrected P values of 3.5×10^{-4}, 9.5×10^{-3}, $< 1.0 \times 10^{-10}$, and 3.2×10^{-4}, and relative risks of 2.50, 2.23, 5.84, and 4.52, respectively.[1] The major studies in this disease are shown in Table 25.

Psoriatic arthropathy is a relatively recently described entity that occurs in approximately 5 to 10% of patients with psoriasis. In this section, we will discuss only the arthritic component of the disease and describe the dermatologic condition and its HLA antigen associations under Dermatology (Chapter 7, Section V).

Typically, psoriatic arthritis is manifested peripherally. In a small group of patients it progresses to a spondylitis, and as it does, the frequency of B27 rises. In Brewerton's study, 46% of the patients with all forms of psoriatic arthritis had the B27 antigen.[49] However, only 24% of those with peripheral disease compared to 68% of those with spinal disease had the B27 antigen. In Metzger's study of 40 patients with psoriatic arthritis, 8 of the 23 (35%) with both peripheral and spinal disease had B27, whereas only 3 of the 17 (18%) with peripheral arthritis alone had B27, for an overall frequency of 28%.[66] In general, psoriatic spondylitis was more often B27-negative than positive. But all the 27-positive spondylitics were males and the duration of psoriasis was 10 years as compared to 18 years in the B27-negative spondylitics. Thus, there may be two forms of spondylitis within the psoriatic group, those with B27 who are males and have a shorter time course and those lacking B27 who are more likely to have an equal sex distribution and have a longer disease history. Although it has been suggested that the presence of B27 does not necessarily predict the development of spondylitis in psoriasis, this finding may be forthcoming as more patients are studied since a similar picture has evolved in juvenile rheumatoid arthritis (described subsequently) in which males with B27 seemed destined to be those who develop spondylitis. In a further clinical breakdown, Brewerton later reported that 10 of 11 psoriatics having both sac-

TABLE 25

Psoriatic Arthropathy

HLA Ag	Site	Controls with Ag (%)	Patients with Ag (%)	Patients with Ag (n)	Patients total (n)	Relative risk	Ref.
B27	Central	4.0	36.4	4	11	12.4	49
	Peripheral	4.0	23.4	11	47	6.5	49
	Central	8.7	16.7	3	18	2.3	42
	Peripheral	8.7	11.6	5	43	1.5	42
	Central	14.1	57.9	11	19	8.2	69
	Peripheral	14.1	23.5	4	17	2.0	69
	Central	3.9	50.0	13	26	22.5	67
	Peripheral	3.9	6.8	3	44	1.9	67
	Central	5.9	34.8	8	23	8.5	66
	Peripheral	5.9	17.6	3	17	3.7	66
B13	Central	4.5	16.7	3	18	4.6	42
	Peripheral	4.5	11.6	5	43	2.9	42
	Central	7.7	36.8	7	19	7.1	69
	Peripheral	7.7	11.8	2	17	1.9	69
	Central	5.3	19.2	5	26	4.3	67
	Peripheral	5.3	9.1	4	44	1.9	67
	Central	3.9	8.7	2	23	2.7	66
	Peripheral	3.9	5.9	1	17	2.1	66
Bw17	Central	4.2	11.1	2	18	3.3	42
	Peripheral	4.2	30.2	13	43	9.8	42
	Central	3.7	10.5	2	19	3.6	69
	Peripheral	3.7	5.9	1	17	2.3	69
	Central	5.9	11.5	3	26	2.2	67
	Peripheral	5.9	22.7	10	44	4.6	67
	Central	9.1	13.0	3	23	1.7	66
	Peripheral	9.1	35.3	6	17	5.6	66
Bw38	Central	2.7	16.7	3	18	7.7	42
	Peripheral	2.7	7.0	3	43	3.0	42
	Central	3.3	26.9	7	26	10.3	67
	Peripheral	3.3	18.2	8	44	6.2	67

roiliitis and spondylitis, 4 of 11 with peripheral arthritis and sacroiliitis without spondylitis, and only 11 out of 47 with psoriasis and peripheral arthritis alone were B27-positive.[49] The increasing frequency of B27 association with psoriasis as one progresses toward the spine is a consistent finding of the Brewerton studies in other patients with inflammatory bowel disease as well as Reiter's disease.

Psoriatic arthritis in the periphery or in the spine may occur several years before skin lesions develop and therefore cannot be considered as a complication of the dermatological condition of psoriasis. This same sequence of events may occur with ulcerative colitis and Crohn's disease, too. It should be noted that there is no increase in B27 in psoriasis uncomplicated by any type of peripheral joint or spine disease, just as in uncomplicated IBD.

The association of the dermatologic condition of psoriasis with antigens B13, Bw17, and Bw16 is described elsewhere, but because of the suggestive evidence that these antigens also show deviations in the different skeletal forms of psoriatic arthritis in 70 patients, namely, B13 with axial disease, Bw17 with peripheral disease, and Bw38 with both types, they are included in Table 23.[67]

The etiology of psoriatic arthritis is even more unclear than the diseases noted above, but it is included in the "seronegative spondarthridities," all of which have certain clinical similarities, suspected but unconfirmed infectious etiologies, and a substantial genetic contribution.[68]

TABLE 26

Juvenile Chronic Polyarthritis

HLA Ag	Site	Controls with Ag (%)	Patients with Ag (%)	Patients with Ag (n)	Patients total (n)	Relative risk	Ref.
B27	Central	7.9	93.3	14	15	10.8	70
	Peripheral	7.9	12.9	4	31	1.5	70
		6.0	42.3	11	26	11.3	71
		8.9	29.2	7	24	4.4	20
		8.9	35.2	19	54	5.5	75
		6.4	14.6	18	123	2.4	72
		10.0	24.2	16	66	2.9	76
		14.1	17.0	8	47	1.3	77
		6.7	57.1	20	35	18.3	45

F. Juvenile Chronic Polyarthritis (JCP)

The primary antigen shown to occur with increased frequency in this disease is B27 (combined uncorrected $P < 1.0 \times 10^{-10}$) with a relative risk (RR) of 4.72 and 95% confidence interval of 3.54 to 6.28.[1] The major studies in this disease are shown in Table 26.

In 1974, again simultaneous studies showed the association of juvenile chronic polyarthritis with B27. In a study of 46 patients by Edmonds, B27 was present in 8 of 8 with sacroiliitis, 6 of 7 with ankylosing spondylitis, 4 of 20 with seronegative juvenile chronic polyarthritis who had not developed sacroiliitis from 11 to 32 years later, and 0 of 11 with persistently seropositive juvenile rheumatoid arthritis with nodules.[70] Because the initial symptom complex of juvenile rheumatoid arthritis may expand with time to include other features such as sacroiliitis without erosion or sclerosis, sacroiliitis with erosion and/or sclerosis, and ankylosing spondylitis at times even with psoriasis or inflammatory bowel disease, these 46 patients were selected with a minimum of a 2-year follow-up from the onset of their disease. Of these 46 patients, the 7 who had presented 15 to 23 years previously with juvenile chronic polyarthritis all subsequently developed definite ankylosing spondylitis and 3 had acute iritis. Furthermore, 8 patients had presented 2 to 11 years previously with juvenile chronic polyarthritis and had developed sacroiliitis of the ankylosing spondylitis type, with 4 developing acute iritis. Of the patients, 20 had been diagnosed as having seronegative juvenile chronic polyarthritis but without sacroiliitis after 11 to 32 years, with 1 having chronic uveitis and 11 having been classified as juvenile rheumatoid arthritis after 9 to 29 years of seronegativity and symmetric joint disease without any evidence of uveitis. In this type of study in which there was, as the authors admit, a bias toward those who had developed at least sacroiliitis, there was not only a high overall frequency of B27 positivity (18 of 46 or 39%), but an extremely high frequency of B27 among those who had developed either sacroiliitis or spondylitis (14 of 15 or 93%). Once again, the heavy male predominance was evident in that all of the spondylitic patients were male, and seven of them had acute iritis in addition.

In Rachelefsky's study of 26 Caucasian patients with juvenile rheumatoid arthritis, 42% had B27 compared to 6% of controls.[71] The clinical characteristics of those with B27 were a larger proportion of males (6 of 11 compared to 3 of 15 without B27), an earlier age of onset (4.5 years compared to 8.4 years), negative serologic tests, and the suggestion that ankylosing spondylitis was more likely to develop in the B27-positive males. The occurrence of uveitis was approximately the same in the B27-positive (1 of 11) as in the B27-negative group (2 of 15).

In a more recent study of 123 patients with juvenile rheumatoid arthritis, the frequency of B27 was surprisingly only 15% compared to 6% of 125 controls.[72] The P value of these data when corrected by multiplying by the number of antigens studied (30) was found to be not significant. Because the authors were specifically looking at the B27 antigen in this disease entity, it seems reasonable that their results would not necessarily have to have been corrected for the 30 antigens studied as was necessary in the initial studies of HLA antigen and disease associations (see Statistical and Genetic Considerations [Chapter 4]). If this possibly unnecessary correction was omitted, the association with B27 became significant. Furthermore, there was a paucity of patients with sacroiliitis in this study, and the presence of ankylosing spondylitis or Reiter's syndrome eliminated any patient from their study. These authors similarly examined a variety of clinical manifestations of the disease and could find no striking relationship for B27 with uveitis, sacroiliitis, tenosynovitis, Sjögren's syndrome, antinuclear antibody titer greater than 1:20, or a positive latex fixation test. The type of onset, presence of a positive family history, nodules, hepatosplenomegaly, lymphadenopathy, pericarditis, rash, fever, and cervical spine involvement showed no statistically significant correlation with B27 either. The occurrence of acute anterior uveitis in this study also differed from that of an earlier study by Brewerton, who reported that 27 of 70 (39%) patients with uveitis but without associated diseases had the B27 antigen.[73]

A study from Belgium of 35 patients with juvenile chronic polyarthritis has shown that B27 occurred in 19 of 20 patients in their two clinical groups composed of JCP evolving to ankylosing spondylitis (N = 3) and JCP with sacroiliitis (N = 17), whereas B27 was present in only 1 of the 15 patients in the remaining two clinical groups who had JCP without sacroiliitis (N = 9) or juvenile rheumatoid arthritis with a positive serology (N = 6).[45]

These studies confirm the observation that the axial or centripetal progression of the disease in seronegative types of arthritis is much more likely to occur in individuals who have the antigen B27. In fact, in juvenile rheumatoid arthritics, those males who have B27 may well be those who will eventually develop sacroiliitis or ankylosing spondylitis.

In a very recent study by Stastny,[74] who divided juvenile rheumatoid arthritis into systemic, polyarticular, and pauciarticular forms, the occurrence of B27 was most prevalent in the pauciarticular form of the disease and particularly in males over the age of 10 years. Only 1 of 23 patients with the systemic form had B27, 1 of 16 with the polyarticular form, and 10 of 48 with the pauciarticular form. When the pauciarticular form is subdivided according to sex and the males again subdivided according to age, it was noted that 5 of 10 males over the age of 10 years and 3 of 11 males under the age of 10 years had B27, whereas only 2 of 27 females had the B27 antigen.

Further work by Stastny has shown that in contrast to the Dw4 antigen that was found to be the predominant D antigen in adult rheumatoid arthritis, the major D-series antigen occurring in juvenile rheumatoid arthritis is one called TMo.[74] This particular antigen was found in 30 of 80 juvenile rheumatoid arthritics (37.5%) compared to only 6% of controls ($P < 0.01$). The other D antigens studied in juvenile rheumatoid arthritis included other private specificities known as JLe, which occurred in 27.2% (22/81) of patients (compared to 9% in controls), and r, which occurred in 14.9% (14/94) of patients (compared to 3% of controls). Dw4 occurred in only 15.5% (15/97) of patients in contrast to 9% of controls.

In order to fully assess the significance and comparative value of these new D-series antigens which have been tested in only one laboratory thus far, it will be necessary to have confirmatory data from other laboratories as well as a comparison of the frequency of these D antigens, particularly TMo, not only in juvenile rheumatoid arthritics as a whole, but also in the subgroups of pauciarticular disease, particularly males

TABLE 27

Adult Rheumatoid Arthritis

HLA Ag	Controls with Ag (%)	Patients with Ag (%)	Patients with Ag (n)	Patients total (n)	Relative risk	Ref.
B27	8.0	8.4	10	119	1.1	2
	5.0	3.8	2	52	0.9	37
	9.5	17.5	7	40	2.1	82
	14.1	48.9	22	45	5.8	83
	9.4	12.0	12	100	1.3	84
	8.4	11.8	19	164	1.4	85

over the age of 10, who have shown such a high frequency of B27 positivity. It may be that the frequency of TMo in these subgroups is even more striking than B27.

G. Rheumatoid Arthritis

Although various reports disclosed slight increases in certain HLA antigens in association with rheumatoid arthritis, no truly significant relationships have been discovered. HLA-A9, B27, Bw40, B13, and Bw21 have all received some attention, but their associations and relative risks are all quite low.[1,78] Data for B27 is given in Table 27 for an appreciation of the low risk associated with B27.

In contrast to the very poor associations of rheumatoid arthritis with any of the A, B, or C series antigens thus far described, there was an impressive association with the D-series antigen Dw4 initially described by Stastny[79] and now confirmed by Svejgaard and Terasaki.[80,81] In adult rheumatoid arthritics, Stastny found that the Dw4 antigen was present in 40% of patients (24/60) compared to only 3% (2 of 77) of controls ($P < 0.01$).[74] In contrast, other D-series antigens that are private to his test panel, such as r, TMo, and JLe, failed to show any such high association and had frequencies in the control population of 3, 6, and 9%, respectively, compared to 15, 33, and 8% in the patient population. The 33% frequency for TMo was based on a study of only 12 patients and was therefore somewhat suspect for its high frequency, although further study of these patients will be necessary.

Thus, the evolving theme of D-locus-associated rheumatic disease is that of female prevalence, seropositivity, and peripheral joint involvement, whereas the theme of B-locus-associated rheumatic diseases such as the B27 cluster of diseases is that of male predominance, axial-oriented disease, and seronegativity.

H. Polymyalgia Rheumatica

In two studies of polymyalgia rheumatica, the first of 15 patients and the second of 50 patients,[86,87] the only common finding was a significant increase in Bw16 or a split of it, Bw38. In the study by Rosenthal, several other antigens were also found to be significantly increased — A10 and Aw32 in the A series and B8, Bw22, and Bw16 in the B series.[86] A different antigen, B5, was found to be increased in the second study, but this was not significant when corrected for the number of antigens studied.[87] The clinical association of polymyalgia rheumatica was also considered by these authors. Rosenthal thought that the association of HLA-A10, an antigen also noted in Hodgkin's disease, suggested a possible interrelationship with neoplasia that can occur in polymyalgia rheumatica.[86] Sany, on the other hand, examined another aspect and suggested that temporal arteritis, which occurred in 19 patients in association with polymyalgia rheumatica, was a distinct disease entity on the basis of an inability to find either B5 or Bw38 in any of 31 patients who had temporal arteritis alone.[87]

I. Gout

In 66 patients with gout studied by Schlosstein as a comparative group in an ankylosing spondylitis study, no HLA antigen association was found.[2]

J. Gonococcal Arthritis

In a study in which gonococcal arthritis was used as a comparative group, no occurrence of B27 was found in 12 such patients, 9 of whom had culture-proven gonococcal arthritis.[47] One small earlier study had shown a surprisingly high frequency with B27 occurring in 5 of 5 cases,[88] but no confirmation was found in the study noted above[47] or in another study with a combined total of 24 patients, not one of whom had B27.[47,89]

K. Vertebral Ankylosing Hyperostosis (Forestier's Disease)

In Forestier's disease, no significant association was found with any HLA antigen in a total of 70 patients studied by Ercilla[90] and Graber-Duvernay,[91] although Shapiro had earlier reported that B27 was significantly increased in 16 of 47 such patients (34%).[92]

L. Paget's Disease

A study conducted of 100 unrelated subjects suffering from obvious Paget's disease, 64 men and 36 women whose average age was 64 years, divided them into two categories according to the number of radiologic lesions: generalized Paget's disease when three or more areas were involved and localized Paget's disease when only one or two were involved.[93] The patients were also considered in another set of categories according to the elevation of the alkaline phosphatase, hydroxyprolinuria, and calcuria. Despite the careful clinical categorization of these 100 patients, no significant differences were noted for any of the 30 HLA specificities tested when compared to 379 unrelated blood donors. Cullen also found no significant HLA antigen differences in a study of 40 patients.[94]

However, very recently, family studies of Paget's disease have disclosed linkage with HLA (lod score 3.157). In each of three informative families the disease segregated with an HLA haplotype (A31/B14, A2/B13, and A11/B18).[95]

M. Frozen Shoulder (Capsulitis, Periarthritis of the Shoulder)

The primary antigen shown to occur with increased frequency in this disease is B27. Bulgen has recently described 38 patients with the frozen shoulder syndrome, 16 of whom (42%) had the antigen B27 with a relative risk of 6.4.[96] This disease, though it might have an immunologic basis, is not known to have any relationship with any other rheumatic diseases, and its association with B27 is unexplained at present. However, in another sense, the association with B27, if confirmed, may help to explain its place in the rheumatic disorders.

N. Idiopathic Scoliosis

A study of 12 unrelated individuals, all Caucasian females, with idiopathic scoliosis showed no significant alteration in the frequency of any HLA antigens, though 7 of the 11 mothers tested had Bw35 and all 4 patients who had the red cell 0 antigen had one of the cross-reacting group, B5, Bw15, B18 and Bw35.[97]

O. Hydroxyapatite Rheumatism

Hydroxyapatite rheumatism, a microcrystalline arthropathy of unknown etiology which may affect single or multiple joints with a calcific periarthritis, was studied in 26 patients, 11 males and 15 females from 15 to 59 years of age. HLA-A2 was found to be present in 19 patients (73% compared to 44% in 591 healthy controls) and Bw35 was found in 10 patients (39% compared to 19% of controls).[98] Furthermore, 6 cases had both the A2 and the Bw35 antigen. Although the uncorrected P values were significant at less than 0.01 and 0.02, respectively, the corrected P values for A2 and Bw35 were not significant.

P. Pyrophosphate Chondrocalcinosis

In 12 patients from 7 families living in an area of Slovakia with an endemic occurrence of this disease, 8 were probably homozygotes for A2 and Bw35 and had a severe clinical course.[99] The A2-Bw35 haplotype occurred in all but 2 of 28 family members, both healthy and with this disease, and the remaining 2 had either A2 or Bw35. In 13 unrelated patients with the same disease but from different areas, the most significant increase in HLA antigen frequencies was still A2 and Bw35.

Q. Alkaptonuria

Alkaptonuria and ochronosis were studied in a 30-member family by Gaucher.[100] In the two members of the first generation, neither was affected and just one had the B27 antigen. Of the second generation, four members were affected, one with ochronotic arthropathy, two with ochronotic spondylosis, and one with alkaptonuria; all four had B27 as did four other unaffected members of this generation.

In 15 of the 21 members of the third generation, B27 was found and occurred in 4 of the 6 members of this generation who had alkaptonuria. In all, 8 of the 10 subjects suffering from alkaptonuria with or without ochronotic arthropathy or spondylosis had the B27 antigen. What the relationship is between the B27 histocompatibility antigen and the enzyme involved in alkaptonuria, homogentisic acid oxidase, is uncertain.

R. Cleft Lip and Cleft Palate

Studies of developmental anomalies were reported by Rapaport, who found five instances of a loss of a parental antigen and three instances of antigen gain in 36 families with cleft lip/cleft palate anomalies. However, only two of these eight abnormal patterns occurred in affected children, the other six being in healthy children.[101] That this might be due to an unequal crossover event was suggested, but the possibility of the extra antigens being B-cell specificities and the deleted antigens representing serologic problems must be considered now also.

In MLC studies, these same authors found that 6 of 21 pairs of unrelated parents appeared to share a D-locus (LD) allele even without SD identical antigens.[102] D-locus typing was unknown at that time, so that the integrity and identity of the postulated allele is uncertain.

S. Spina Bifida

Spina bifida was studied in families by Bobrow, who sought to find a human counterpart to the H-2 linked T locus that is responsible for abnormalities of the notochord and axial skeleton in mice.[103] These studies have had negative results thus far. However, Amos' study of spina bifida occulta and asymmetry of the facet joints in a single large family revealed an association between these abnormalities and the ancestral haplotypes.[104]

T. Rheumatic Fever (see also Infection and Immunization [Chapter 7, Section VIII])

In studies by Caughey[105] and Falk[106] rheumatic fever including rheumatic heart disease was reported to show no significant increase in any HLA antigen. However, in these two studies, there was a decrease in A3 that was found to be even more striking in a preliminary study of 48 patients from Mexico.[107] But this latter series was too small to establish the point and no control population data were provided. However, Ward found a significant increase in Aw30/31 and A29 in those with rheumatic heart disease who had no rheumatic history.[108] No confirmation was found for Falk's finding that there was an increase in homozygotes.[106]

U. Unclassified Inflammatory Rheumatic Diseases (UIRD)

A study by Amor of 115 patients with unclassified inflammatory rheumatic diseases showed the B27 antigen to be present in 45% of the 80 males and 37% of the 35 females.[109] There was no difference in the sex distribution of the B27 antigen or of the age of onset (before or after the age of 40 years). Furthermore, there were no differences according to the site of involvement: spinal — 42% B27-positive; peripheral joints — 37% B27-positive; and mixed forms — 51% B27-positive. However, iritis, which occurred in 14 patients, was accompanied by B27 in 10 cases. The authors concluded that the group of UIRD is predominantly composed of male patients (70%) whose disease usually begins before the age of 40 (83.4%) with a percentage of B27 positivity of 42.6%, intermediate between ankylosing spondylitis at one extreme and rheumatoid arthritis at the other. These patients were clinically felt to have atypical AS or Reiter's syndrome.

V. Postviral Arthropathy

A low frequency of B27 was found in both presumed rubella arthritis (13%; 3 of 23 cases) and mixed viral arthritis (29%; 5 of 17 cases).[110]

REFERENCES

1. **Ryder, L. P. and Svejgaard, A.**, Associations Between HLA and Disease, Report from the HLA and Disease Registry of Copenhagen, 1976.
2. **Schlosstein, L., Terasaki, P. I., Bluestone, R., and Pearson, C. M.**, High association of an HL-A antigen, W27, with ankylosing spondylitis, *N. Engl. J. Med.*, 288, 704, 1973.
3. **Brewerton, D. A., Hart, F. D., Nicholls, A., Caffrey, M., James, D. C. O., and Sturrock, R. D.**, Ankylosing spondylitis and HL-A27, *Lancet*, 1, 904, 1973.
4. **Khan, M. A., Braun, W. E., and Kushner, I.**, Low frequency of HLA-B27 in American Blacks with ankylosing spondylitis, *Clin. Res.*, 24, 331A , 1976.
5. **Lockshin, M. D., Fotino, M., Gough, W. W., and Litwin, S. D.**, Ankylosing spondylitis and HL-A; a genetic disease plus?, *Am. J. Med.*, 58, 695, 1975.
6. **Levitin, P. M., Gough, W. W., and Davis, J. S., IV**, HLA-B27 antigen in women with ankylosing spondylitis, *JAMA*, 235, 2621, 1976.
7. **Swezey, R. L., Zucker, L. M., and Terasaki, P. I.**, Reduced prevalence of HL-A antigen W27 in Black females with ankylosing spondylitis, *J. Rheumatol.*, 1, 260, 1974.
8. **Good, A. E., Kawanishi, H., and Schultz, J. S.**, HLA B27 in Blacks with ankylosing spondylitis or Reiter's disease, *N. Engl. J. Med.*, 294, 166, 1976.
9. **Espinoza, L., Oh, J. H., Kinsella, T. D., Stacey, C. H., Osterland, C. K., and Dove, F. B.**, Anklyosing spondylitis: Family studies and HL-A27 antigen distribution, *J. Rheumatol.*, 1, 254, 1974.
10. **Myers, B. W. and Cheatum, D. E.**, Ankylosing spondylitis in a Black woman, *JAMA*, 231, 278, 1975.
11. **Mills, D. M., Arai, Y., and Gupta, R. C.**, HL-A antigens and sacroiliitis, *JAMA*, 231, 268, 1975.
12. **Botha, M. C.**, personal communication, 1976.
13. **Gofton, J. P., Chalmers, A., Price, G. E., and Reeve, C. E.**, HL-A27 and ankylosing spondylitis in B.C. Indians, *J. Rheumatol.*, 2, 314, 1975.
14. **Calin, A., Bennett, P. H., Jupiter, J., and Terasaki, P. I.**, HLA-B27 and sacroiliitis in Pima Indians, in *HLA and Disease*, INSERM, Paris, 1976, 18.
15. **Sonozaki, H., Seki, H., Chang, S. Okuyama, M., and Juji, T.**, Human lymphocyte antigen, HL-A27, in Japanese patients with ankylosing spondylitis, *Tissue Antigens*, 5, 131, 1975.

16. **Simons, M. J.**, personal communication, 1976.
17. **Brautbar, C., Porath, S., Nelken, D., and Cohen, T.**, HLA-B27 and ankylosing spondylitis in the Israeli population, in *HLA and Disease,* INSERM, Paris, 1976, 17.
18. **Contu, L., Capelli, G., and Sale, S.**, HLA-B27 and ankylosing spondylitis (AS): A population and family study in Sardinia, in *HLA and Disease,* INSERM, Paris, 1976, 24.
19. **van der Linden, J. M., Keuning, J. J., Wuisman, J. H., Cats, A., and van Rood, J. J.**, HL-A27 and ankylosing spondylitis, *Lancet,* 1, 520, 1975.
20. **Sturrock, R. D., Dick, H. M., Henderson, N., Goel, G. K., Lee, P., Dick, C., and Buchanan, W. W.**, Association of HL-A27 and AJ in juvenile rheumatoid arthritis and ankylosing spondylitis, *J. Rheumatol.,* 1, 269, 1974.
21. **Woodrow, J. C. and Eastmond, C. J.**, HLA B27 and the genetics of ankylosing spondylitis (A.S.), in *HLA and Disease,* INSERM, Paris, 1976, 56.
22. **de Bruyere, M. and de Deuxchaisnes, Ch. N.**, Segregation of HL-A27 and ankylosing spondylitis in an informative kindred, *Tissue Antigens,* 7, 15, 1976.
23. **Jeannet, M., Saudan, Y., and Bitter, T.**, HL-A27 in female patients with ankylosing spondylitis, *Tissue Antigens,* 6, 262, 1975.
24. **Möller, E. and Olhagen, B.**, Studies on the major histocompatibility system in patients with ankylosing spondylitis, *Tissue Antigens,* 6, 237, 1975.
25. **Svejgaard, A., Platz, P., Ryder, L. P., Staub-Nielsen, L., and Thomsen, M.**, HL-A and disease associations: a survey, *Transplant. Rev.,* 22, 3, 1975.
26. **Calin, A., Fries, J. F., Stinson, F. B., and Payne, R.**, Normal frequency of HLA B27 in aortic insufficiency, in *HLA and Disease,* INSERM, Paris, 1976, 21.
27. **Calin, A. and Fries, J. F.**, Striking prevalence of ankylosing spondylitis in "healthy" W27 positive males and females, *N. Engl. J. Med.,* 293, 835, 1975.
28. **Cohen, L. M., Mittal, K. K., Schmid, F. R., Rogers, L. F., and Cohen, K. L.**, Increased risk for spondylitis stigmata in apparently healthy HL-AW27 men, *Ann. Intern. Med.,* 84, 1, 1976.
29, **Albert, E.**, personal communication, 1976.
30. **Terasaki, P. I. and Mickey, M. R.**, HL-A haplotypes of 32 diseases, *Transplant. Rev.,* 22, 105, 1975.
31. **Kastelan, A., Kerhin-Brkljacic, V., Jajic, I., Brkljacic, L., and Balog, V.**, The frequency of HL-A haplotypes and their segregation analysis in families of patients with ankylosing spondylitis, in *HLA and Disease,* INSERM, Paris, 1976, 39.
32. **Dausset, J. and Hors, J.**, Some contributions of the HL-A complex to the genetics of human disease, *Transplant. Rev.,* 22, 44, 1975.
33. **Sachs, J. A., Sterioff, S., Robinette, M., Wolf, E., Curry, H. L., and Festenstein, H.**, Ankylosing spondylitis and the major histocompatibility locus, *Tissue Antigens,* 5, 120, 1975.
34. **Ebringer, A., Cowling, P., Ngwa Suh, N., James, D. C. O., and Ebringer, R. W.**, Crossreactivity between Klebsiella aerogenes species and B27 lymphocyte antigens as an aetiological factor in ankylosing spondylitis, in *HLA and Disease,* INSERM, Paris, 1976, 27.
35. **Brewerton, D. A.**, Rheumatology Workshop Summary, First Int. Symp. on HLA and Disease, Paris, June 23, 1976.
36. **Amor, B., Feldmann, J.-L., Delbarre, F., Hors, J., Beaujan, M. M., and Dausset, J.**, L'antigène HL-A W27: sa fréquence dans la spondylarthrite ankylosante et le syndrome de Fiessinger-Leroy-Reiter, *Nouv. Presse Med.,* 3, 1373, 1974.
37. **Marcolongo, R. and Contu, L.**, Les antigènes HL-A dans polyarthrite chronique rhumatismale et la spondylarthrite ankylosante en Sardaigne, *Nouv. Presse Med.,* 3, 2023, 1974.
38. **Russell, A. S., Schlaut, J., Percy, J. S., and Dossetor, J. B.**, HL-A (transplantation) antigens in ankylosing spondylitis and Crohn's disease, *J. Rheumatol.,* 1, 203, 1974.
39. **Calin, A., Grahame, R., Tudor, M., and Kennedy, L.**, "Ankylosing rheumatoid arthritis," ankylosing spondylitis, and HL-A antigens, *Lancet,* 1, 874, 1974.
40. **Dick, H. M., Sturrock, R. D., Goel, G. K., Henderson, N., Canesi, B., Rooney, P. J., Dick, W. C., and Buchanan, W. W.**, The association between HL-A antigens, ankylosing spondylitis, and sacro-iliitis, *Tissue Antigens,* 5, 26, 1975.
41. **Cross, R. A., Rigby, R., and Dawkins, R. L.**, The significance of HL-A W27 in ankylosing spondylitis and Reiter's syndrome with three family studies, *Aust. N. Z. J. Med.,* 5, 108, 1975.
42. **Sany, J., Seignalet, J., Guilhon, J.-J., and Serre, H.**, HL-A et rhumatisme psoriasique, *Rev. Rhum. Mal. Osteo. Articulaires,* 42, 451, 1975.
43. **Truog, P., Steiger, U., Contu, L., Galfré, G., Trucco, M., Bernoco, D., Bernoco, M., Birgen, I., Dolivo, P., and Ceppellini, R.**, Ankylosing spondylitis (AS): A population and family study using HL-A serology and MLR, in *Histocompatibility Testing 1975,* Kissmeyer-Nielsen, F., Ed., Munksgaard, Copenhagen, 1975, 788.
44. **van Rood, J. J., van Hooff, J. P., and Keuning, J. J.**, Disease predisposition, immune responsiveness and the fine structure of the HL-A supergene, *Transplant. Rev.,* 22, 75, 1975.
45. **Veys, E. M., Coigne, E., Mielants, H., and Verbruggen, A.**, HL-A and juvenile chronic polyarthritis, *Tissue Antigens,* 8, 61, 1976.

46. **Brewerton, D. A., Nicholls, A., Oates, J. K., Caffrey, M., Walters, D., and James, D. C. O.,** Reiter's disease and HL-A27, *Lancet,* 2, 996, 1973.
47. **Morris, R., Metzger, A. L., Bluestone, R., and Terasaki, P. I.,** HL-A W27 — a clue to the diagnosis and pathogenesis of Reiter's syndrome, *N. Engl. J. Med.,* 290, 554, 1974.
48. **McClusky, O. E., Lordon, R. E., and Arnett, F. C.,** HL-A27 in Reiter's syndrome and psoriatic arthritis: a genetic factor in disease susceptibility and expression, *J. Rheumatol.,* 1, 263, 1974.
49. **Brewerton, D. A. and James, D. C. O.,** The histocompatibility antigen (HL-A27) and disease, *Semin. Arthritis Rheum.,* 4, 191, 1975.
50. **Arnett, F. C., McClusky, O. E., Schacter, B. Z., and Lordon, R. E.,** Incomplete Reiter's syndrome: discriminating features and HL-A W27 in diagnosis, *Ann. Intern. Med.,* 84, 8, 1976.
51. **Woodrow, J. C., Treanor, B., and Usher, N.,** The HL-A system in Reiter's syndrome, *Tissue Antigens,* 4, 533, 1974.
52. **Aho, K., Ahvonen, P., Lassus, A., Sievers, K., and Tiilikainen, A.,** HL-A27 in reactive arthritis: a study of yersinia arthritis and Reiter's disease, *Arthritis Rheum.,* 17, 521, 1974.
53. **Zachariae, H., Friis, J., Graudal, H., Hjortshøj, A., Kissmeyer-Nielsen, F., Svejgaard, A., Svejgaard, E., and Zachariae, E.,** Reiter's disease and the histocompatibility antigen HL-A27, *Scand. J. Rheumatol.,* 4, 13, 1975.
54. **Paronen, I.,** Reiter's disease: A study of 344 cases in Finland, *Acta Med. Scand. Suppl.,* 131 (Suppl. 212), 1, 1948.
55. **Sairanen, E., Paronen, I., and Mahonen, H.,** Reiter's syndrome: A follow-up study, *Acta Med. Scand.,* 185, 57, 1969.
56. **Sairanen, E. and Tiilikainen, A.,** HL-A27 in Reiter's disease following shigellosis, Rheumatology Workshop, 1st Int. Symp. on HLA and Disease, Paris, June 1976.
57. **Calin, A. and Fries, J. F.,** An "experimental" epidemic of Reiter's syndrome revisited, *Ann. Intern. Med.,* 84, 564, 1976.
58. **Aho, K., Ahvonen, P., Lassus, A., Sievers, K., and Tiilikainen, A.,** cited in **Brewerton, D. A. and James, D. C.,** The histocompatibility antigen (HL-A27) and disease, *Semin. Arthritis Rheum.,* 4, 191, 1975.
59. **Hakansson, U., Low, B., Eitrem, R., and Winblad, S.,** HL-A27 and reactive arthritis in an outbreak of Salmonellosis, *Tissue Antigens,* 6, 366, 1975.
60. **Leirisalo, M., Tiilikainen, A., and Laitinen, O.,** HLA phenotypes in rheumatic fever and Yersinia arthritis, in *HLA and Disease,* INSERM, Paris, 1976, 41.
61. **Friis, J. and Svejgaard, A.,** Salmonella arthritis and HL-A27, *Lancet,* 1, 1350, 1974.
62. **Bluestone, R.,** HL-A W27 and the "rheumatoid variants," *Hosp. Pract.,* 10, 131, 1975.
63. **Macrae, I. and Wright, V.,** A family study of ulcerative colitis, with particular reference to ankylosing spondylitis and sacroiliitis, *Ann. Rheum. Dis.,* 32, 16, 1973.
64. **Morris, R. I., Metzger, A. L., Bluestone, R., and Terasaki, P. I.,** HL-A-W27 — A useful discriminator in the arthropathies of inflammatory bowel disease, *N. Engl. J. Med.,* 290, 1117, 1974.
65. **Nagant de Deuxchaisnes, C., Haux, J. P., Fiasse, R., and de Bruyere, M.,** Ankylosing spondylitis, sacroiliitis, regional enteritis, and HL-A27, *Lancet,* 1, 1238, 1974.
66. **Metzger, A. L., Morris, R. I., Bluestone, R., and Terasaki, P. I.,** HL-AW27 in psoriatic arthropathy, *Arthritis Rheum.,* 18, 111, 1975.
67. **Feldmann, J. L., Amor, B., Cazalis, P., Dryll, A., Hors, J., and Hacquart, B.,** Antigènes HLA chez les malades atteints de rheumatisme psoriasique, *Nouv. Presse Med.,* 5, 477, 1976.
68. **Moll, J. M. H. and Wright, V.,** Familial occurrence of psoriatic arthritis, *Ann. Rheum. Dis.,* 32, 181, 1973.
69. **Karvonen, J., Tiilikainen, A., and Lassus, A.,** HL-A antigens in patients with persistent palmoplantar pustulosis and pustular psoriasis, *Ann. Clin. Res.,* 7, 112, 1975.
70. **Edmonds, J., Morris, R. I., Metzger, A. L., Bluestone, R., Terasaki, P. I., Ansell, B., and Bywaters, E. G. L.,** Follow-up study of juvenile chronic polyarthritis with particular reference to histocompatibility antigen W27, *Ann. Rheum. Dis.,* 33, 289, 1974.
71. **Rachelefsky, G. S., Terasaki, P. I., Katz, R., and Stiehm, E. R.,** Increased prevalence of W27 in juvenile rheumatoid arthritis, *N. Engl. J. Med.,* 290, 892, 1974.
72. **Gibson, D. J., Carpenter, C. B., Stillman, S., and Schur, P. H.,** Re-examination of histocompatibility antigens found in patients with juvenile rheumatoid arthritis, *N. Engl. J. Med.,* 293, 636, 1975.
73. **Brewerton, D. A., Caffrey, M., Nicholls, A., Walters, D., and James, D. C.,** Acute anterior uveitis and HL-A27, *Lancet,* 1, 464, 1974.
74. **Stastny, P.,** personal communication, 1976.
75. **Buc, M., Nyulassy, S., Stefanovic, J., Michalko, J., and Mozolova, D.,** HLA system and juvenile rheumatoid arthritis, *Tissue Antigens,* 4, 395, 1974.
76. **Hall, M. A., Ansell, B. M., James, D. C., and Zylinski, P.,** HL-A antigens in juvenile chronic polyarthritis (Still's disease), *Ann. Rheum. Dis.,* 34 (Suppl. 1), 36, 1975.
77. **Nissila, M., Elomaa, L., and Tiilikainen, A.,** HL-A antigens in juvenile rheumatoid arthritis, *N. Engl. J. Med.,* 292, 430, 1975.

78. **Seignalet, J., Clot, J., Sany, J., and Serre, H.,** HL-A antigens in rheumatoid arthritis, *Vox Sang.,* 23, 468, 1972.
79. **Stastny, P.,** Mixed lymphocyte culture typing cells from patients with rheumatoid arthritis, *Tissue Antigens,* 4, 571, 1974.
80. **Svejgaard, A.,** personal communication, 1976.
81. **Terasaki, P. I.,** personal communication, 1976.
82. **Nyulassy, S., Ravingerova, G., Zvarova, E., and Buc, M.,** HL-A antigens in rheumatoid arthritis, *Lancet,* 1, 450, 1974.
83. **Isomaki, H., Koota, K., Maitio, J., Nissila, M., and Tiilikainen, A.,** HL-A27 and arthritis, *Ann. Clin. Res.,* 7, 138, 1975.
84. **Seignalet, J.,** cited in **Ryder, L. P. and Svejgaard, A.,** Associations Between HLA and Disease: Report from the HLA and Disease Registry of Copenhagen, 1976.
85. **Huaux, J. P., Devogelaer, J. P., de Bruyere M., and Nagant de Deuxchaisnes, C.,** "Artificial" elevation of HLA-B27 incidence in control groups, versus a residual significant elevation in rheumatoid arthritis, in *HLA and Disease,* INSERM, Paris, 1976, 304.
86. **Rosenthal, M., Muller, W., Albert, E. D., and Schattenkirchner, M.,** HL-A antigens in polymyalgia rheumatica, *N. Engl. J. Med.,* 292, 595, 1975.
87. **Sany, J., Seignalet, J., Serre, H., and Lapinski, H.,** HLA in polymyalgia rheumatica, in *HLA and Disease,* INSERM, Paris, 1976, 49.
88. **Aho, K., Ahvonen, P., Lassus, A., Sievers, K., and Tiilikainen, A.,** HL-A antigen 27 and reactive arthritis, *Lancet,* 2, 157, 1973.
89. **Wagner, L. P. and Fessel, W. J.,** HL-A27 (W27) absent in gonococcal arthritis, *Lancet,* 1, 1094, 1975.
90. **Ercilla, G., Brancós, M. A., Breysse, Y., Vives, J., Rotés, J., Castillo, R., and Alonso, G.,** Histocompatibility antigens in Forestier's disease, polyarthrosis and ankylosing spondylitis, in *HLA and Disease,* INSERM, Paris, 1976, 28.
91. **Graber Duvernay, B., Gras, J. P., Dutertre, P., and Bensa, J. C.,** Histocompatibility antigens in Forestier's disease, in *HLA and Disease,* INSERM, Paris, 1976, 35.
92. **Shapiro, R. F., Wiesner, K. B., Bryan, B. L., Utsinger, P. D., Resnick, D., and Castles, J. J.,** HLA-B27 and modified bone formation, *Lancet,* 1, 230, 1976.
93. **Mercier, P., Roux, H., Maestracci, D., Serratrice, G., Seignalet, J., Simon, L., and Blotman, F.,** Paget's disease and H.L.A. antigens, in *HLA and Disease,* INSERM, 1976, 44.
94. **Cullen, P., Russell, R. G. G., Walton, R. J., and Whiteley, J.,** Frequencies of HLA-A and HLA-B histocompatibility antigens in Paget's disease of bone, *Tissue Antigens,* 7, 55, 1976.
95. **Fotino, M. and Haymovitz, A.,** Evidence for linkage between HLA and Paget's disease, *Transplant. Proc.,* in press.
96. **Bulgen, D. Y., Hazleman, B. L., and Voak, D.,** HLA-B7 and frozen shoulder, *Lancet,* 1, 1042, 1976.
97. **Braun, W. E., Zachary, A., Rothner, A. D., and Nash, C. L.,** Idiopathic scoliosis and HLA antigens, Copenhagen Registry, submitted.
98. **Amor, B., Cherot, A., Delbarre, F., Nunez-Roldan, A., Hors, J., and Dausset, J.,** Hydroxyapatite rheumatism and HLA, in *HLA and Disease,* INSERM, Paris, 1976, 13.
99. **Nyulassy, S., Stefanovic, J., and Sitaj, S.,** HL-A homozygosity and calcium pyrophosphate metabolism. Preliminary report, in *Histocompatibility Testing 1975,* Kissmeyer-Nielsen, F., Ed., Munksgaard, Copenhagen, 1975, 805.
100. **Gaucher, A., Raffoux, C., Netter, P., Faure, G., Janot, J., Pourel, J., and Streiff, F.,** HLA antigens and alcaptonuria, in *HLA and Disease,* INSERM, Paris, 1976, 31.
101. **Rapaport, F. T., Bachvaroff, R., Converse, J. M., Raisbeck, A. P., Ayrazian, J. H., Segall, M., and Bach, F. H.,** Atypical patterns of inheritance of the serologically detectable (SD) products of the HL-A complex in human developmental anomalies, *Transplant. Proc.,* 5, 1817, 1973.
102. **Rapaport, F. T., Bachvaroff, R., Converse, J. M., Raisbeck, A. P., Ayrazian, J. H., Segall, M., and Bach, F. H.,** Genetic studies of mixed leukocyte culture reactivity in human developmental anomalies, *Transplant. Proc.,* 5, 1823, 1973.
103. **Bobrow, M., Bodmer, J. G., Bodmer, W. F., McDevitt, H. O., Lorber, J., and Swift, P.,** The search for a human equivalent of the mouse T-locus-negative results from a study of HL-A types in spina bifida, *Tissue Antigens,* 5, 234, 1975.
104. **Amos, D. B., Ruderman, R., Mendell, N. R., and Johnson, A. H.,** Linkage between HL-A and spinal development, *Transplant. Proc.,* 7 (Suppl 1), 93, 1975.
105. **Caughey, D. E., Douglas, R., Wilson, W., and Hassall, I. B.,** HL-A antigens in Europeans and Maoris with rheumatic fever and rheumatic heart disease, *J. Rheumatol.,* 2, 319, 1975.
106. **Falk, J. A., Fleischman, J. L., Zabriskie, J. B., and Falk, R. E.,** A study of HL-A antigen phenotype in rheumatic fever and rheumatic heart disease patients, *Tissue Antigens,* 3, 173, 1973.
107. **Gorodezky, C.,** HLA and rheumatic heart disease, in *HLA and Disease,* INSERM, Paris, 1976, 34.

108. **Ward, C., Gelsthorpe, K., Doughty, R. W., and Hardisty, C. A.,** HLA antigens and acquired valvular heart disease, *Tissue Antigens,* 7, 227, 1976.
109. **Amor, B., Kahan, A., and Delbarre, F.,** HLA-B27 in unclassified inflammatory rheumatic diseases (UIRD), in *HLA and Disease,* INSERM, Paris, 1976, 14.
110. **Robitaille, A., Cockburn, C., James, D. C., and Ansell, B. M.,** HLA frequencies in less common arthropathies, *Ann. Rheum. Dis.,* 35, 271, 1976.

XV. MISCELLANEOUS DISEASE

A. Hereditary Hemorrhagic Telangiectasia

A study of 20 individuals from three generations of the same family, 5 with hereditary hemorrhagic telangiectasia, revealed that the haplotype A2-Bw17 occurred in all with the disease.[1] This same haplotype was not found in the clinically healthy members with the exception of two persons in the second generation.

B. Von Willebrand's Disease

The frequencies of B7 and Bw35 were increased in 34 patients to 41.2 and 35.3%, respectively, compared to 26.9 and 17.2%, respectively, in controls.[2] Another study of 28 unrelated patients with the variable form of Willebrand's disease showed a decreased frequency of A28 and B7 to 3.5 and 7.1%, respectively, compared to 26.6 and 24.2%, respectively, in controls.[3] Neither of these differences was significant after correction.[3] Small increases were also noted in A29 and Aw32, B12, and Bw17, but again these differences were not significant.[3] Segregation analyses among 26 families suggested that these latter four antigens were preferentially inherited by the patients, whereas A28 and B7 appeared to be transmitted more to the unaffected siblings. These two studies presented conflicting data on B7 and offered no immediate basis for resolving the differences.

C. Autoimmune Hemolytic Anemia (Idiopathic)

In a single study, 36 cases had no significant alteration in HLA antigen frequency.[4]

D. Idiopathic Thrombocytopenic Purpura

No significant HLA antigen frequency alterations were found in 48 cases.[4]

E. Acute Intermittent Porphyria

In 20 patients with acute intermittent porphyria (AIP), all apparently unrelated, there were increases in A1, A10, A13, and Aw29, and decreases in A2, A3, A9, and A11.[5] All of these differences were only of borderline significance.

F. Glucose 6 Phosphate Dehydrogenase (G6PD) Deficiency

This congenital condition that predisposes to hemolytic anemia was studied in 450 subjects, 100 of whom had a total congenital G6PD deficiency. There were increases in A10, B12, and Bw40, but specific control data were lacking.[6]

G. Familial Mediterranean Fever

A single study showed that 31 patients, 15 to 54 years of age, all of Sephardic Israelite origin except one, showed no significant HLA antigen differences from a control population of Jewish Yemenites.[7] Only an increased frequency of A3 (29% compared to 8% of controls) was noted. In seven families with at least two siblings affected, no relationship could be found between the disease and any specific HLA haplotype.

H. Hurler's Disease

Of eight genotyped patients with this type of mucopolysaccharidosis, four had an A2-B5 haplotype.[4]

I. Down's Syndrome

The fact that 76 patients studied by Segal had no significant alteration in HLA antigen frequency suggested the chromosome 21 had essentially no influence on HLA expression.[8] The normal antigen frequency in patients implied normal antigen frequency in their parents so that the development of Down's syndrome could not be predicted in this way.

J. Azoospermy

An increase in A28 was found in 39 azoospermic patients without structural defects (21% compared to 6% of controls) which was of borderline significance (corrected P < 0.03). A nonsignificant increase in Bw40 was also found (21% compared to 10% of controls).[9]

K. Preeclampsia and Eclampsia

An increased frequency of A1 and a decreased frequency of A9, B14, and Bw17 were found among 37 women with severe preeclampsia. Interestingly, none of these women produced anti-HLA antibodies specific for their husbands' lymphocytes during or after their pregnancy.[10] However, in 46 women with either eclampsia or preeclampsia, Scott found no significant alteration in any HLA antigen frequency,[11] although A9 and B14 were decreased as reported by Jenkins.[10]

L. Triploid Conceptus

Because of the possibility that HLA antigens may be an important factor at the time of fertilization, these antigens were investigated in 49 couples who had had a spontaneous abortion of a triploid conceptus.[12] However, no significant aberration of HLA antigen frequency was found. Only a small increase in the frequency of shared parental antigens was found in nine couples (18% compared to 11% in normals). Such a parental sharing of antigens is the same type of background necessary for the homozygosities seen in children with certain leukemias and individuals with aplastic anemia.[13]

M. Aging

Aging may be considered a disease whose possible association with HLA antigens might be apparent by comparing the HLA antigens of the very young and very old. In 1976, Bender terated a simple model proposed by him in 1973 in which a high degree of heterozygosity would provide better immunologic surveillance and afford a longer life expectancy.[14] Although one could readily defend the opposite point — that it is easier to detect change on a simpler background (i.e., viral or tumor antigens appearing on a homozygous background rather than on a more complex heterozygous background) — experiments by Doherty, Blanden, and Zinkernagel on the H-2 restriction of cytotoxicity for virus-infected cells constitute an elegant argument in favor of heterozygosity being a protective mechanism for survival.[15] Clinical support for this concept was provided by Gerkins' data that failed to show even one double homozygosity in 126 healthy persons more than 75 years of age.[16] Similarly, Macurova found that two B-series antigens occurred in 72% of the aged ($>$ 80 years) and only 54% of young people ($<$ 20 years).[17]

Conflicting results also have been found. A significant increase in Bw40 (corrected $P < 0.02$),[17] a prevalence of A1 and B8,[16] and no alteration in HLA antigen freqaency[14] have all been reported in the aged. The recent data of Bender that showed no difference in the number of people with four, three, or two antigens in the young- or old-age groups are further confusing the issue (66.7, 30.0, and 3.3% compared to 63.3, 32.0, and 4.7%, respectively). Thus, there appears to be no strong indication at this time of greater HLA heterozygosity or of a particular HLA antigen prevalence that might indicate a survival characteristic expressed in the aged.

N. Polycythemia Vera

The largest study of polycythemia vera showed that in 122 patients there was a non-significant increase in Bw21 (17% compared to 7.5% of controls) and a moderate decrease in B13 (3% compared to 8.5% of controls).[18]

REFERENCES

1. **Kissel, P., Raffoux, C., Andre, J.M., Faure, G., Netter, P., and Streiff, F.**, HLA antigens and hereditary hemorrhagic telangiectasia, in *HLA and Disease,* INSERM, Paris, 1976, 253.
2. **Müller, N., Budde, U., and Etzel, F.**, von Willebrand's disease and HLA-antigens, in *HLA and Disease,* INSERM, Paris, 1976, 258.
3. **Goudemand, J., Mazurier, C., and Parquet-Gernez, A.**, HLA antigens and Willebrand's disease, in *HLA and Disease,* INSERM, Paris, 1976, 247.
4. **Dausset, J. and Hors, J.**, Contributions of the HL-A complex to the genetics of human diseases, *Transplant. Rev.,* 22, 44, 1975.
5. **Teuber, J., Druschky, F., and Baenkler, H.W.**, HLA and acute intermittent porphyria, in *HLA and Disease,* INSERM, Paris, 1976, 265.
6. **Carcassi, U., Del Giacco, G.S., Pintus, A., Perpignano, G., Locci, F., Loy, M., Piludu, G., and Leone, A.L.**, HLA antigens and erythrocyte glucose-6-phosphate dehydrogenase (G-6-PD) deficiency in Sardinia, in *HLA and Disease,* INSERM, Paris, 1976, 243.
7. **Chaouat, Y., Tormen, J.P., Godeau, P., Kahn, M.F., Ryckewaert, A., Schmid, M., and Hors, J.**, HLA markers and periodic disease (familial Mediterranean fever) (FMF), in *HLA and Disease,* INSERM, Paris, 1976, 244.
8. **Segal, D.J., Schlaut, J.W., Pabst, H.F., McCoy, E.E., and Dossetor, J.B.**, HL-A frequencies in Down's syndrome, *Humangenetik,* 27, 45, 1975.
9. **Bisson, J.P., David, G., Kolevski, P., Hors, J., and Dausset, J.**, Azoospermy and HLA antigens, in *HLA and Disease,* INSERM, Paris, 1976, 242.
10. **Jenkins, D.M., Need, J.A., Scott, J.S., and Rajah, S.M.**, HLA and severe pre-eclampsia, in *HLA and Disease,* INSERM, Paris, 1976, 250.
11. **Scott, J.R., Beer, A.E., and Stastny, P.**, Immunogenetic factors in pre-eclampsia and eclampsia, *JAMA,* 235, 402, 1976.
12. **Couillin, P., Boue, A., Boue, N., Ravise, N., and Hors, J.**, HLA markers in parents of triploid conceptuses, in *HLA and Disease,* INSERM, Paris, 1976, 245.
13. **Gluckman, E., Lemarchand, F., Nunez-Roldan, A., Hors, J., and Dausset, J.**, Possible excess of HLA-A homozygous among aplastic anemia and acute lymphoblastic leukemia (ALL), in *HLA and Disease,* INSERM, Paris, 1976, 226.
14. **Bender, K., Mayerova, A., Klotzbucher, B., Burckhardt, K., and Hiller, Ch.**, No indication of post-natal selection at the HL-A loci, *Tissue Antigens,* 7, 118, 1976.
15. **Doherty, P.C., Blanden, R.V., and Zinkernagel, R.M.**, Specificity of virus-immune effector T cells for H-2K or H-2D compatible interactions: implications for H-antigen diversity, *Transplant. Rev.,* 29, 89, 1976.
16. **Gerkins, V.R., Ting, A., Menck, H.T., Casagrande, J.T., Terasaki, P.I., Pike, M.C., and Henderson, B.E.**, HL-A heterozygosity as a marker of long term survival, *J. Natl. Cancer Inst.,* 52, 1909, 1974.
17. **Macurova, H., Ivanyi, P., Sajdlova, H., and Trojan, J.**, HL-A antigens in aged persons, *Tissue Antigens,* 6, 269, 1975.
18. **Benjamin, D., Zamir, R., Gazit, E., Pinkhas, J., and Yehoshua, H.**, HLA antigens in polycythemia vera, *Tissue Antigens,* 11, 65, 1978.

Chapter 8

CONCLUDING REMARKS

It is clear that the vast majority of HLA studies performed in a tremendous variety of diseases has not yielded any clinically significant association with HLA antigens or any evidence of linkage with HLA loci. However, the dramatic finding of a strong association between HLA-B27 and both ankylosing spondylitis and Reiter's disease indicates the impressive and clinically useful information that potentially can be derived from such studies. Furthermore, the finding of linkage between HLA antigens and diseases such as Paget's disease as well as a form of spinocerebellar ataxia permits the identification of individuals at risk for these diseases within families.

The general applicability of HLA association or HLA linkage as a means of identifying high-risk patients for certain diseases may expose more basic pathogenetic mechanisms for these diseases. It is primarily this concept rather than the few clinically useful disease associations or disease linkages which has made this area one of the most exciting in medicine today. Moreover, the rapid developments in identifying immune response genes in various animal species signal a strong likelihood of a similar system or systems existing in man also. Studies are already being conducted in this area under the designation of B-cell, Ia, or DR antigen studies. It may be that as the genetics of the major histocompatibility complex unfold in animals and man, a remarkably similar spectrum of genetic and immunologic relationships will be revealed that govern the entire animal kingdom and man himself.

INDEX

A

B

C

D

E

F

G

H

I

J

K

L

M

N

O

P

R

S

T

U

V

W

X

Y